OCR

Physics

REVISION GUIDE

Stephen Pople
Carol Tear

A2

OXFORD
UNIVERSITY PRESS

Oxford University Press is a department of the University of Oxford. It furthers the University's objective of excellence in research, scholarship, and education by publishing worldwide in

Oxford New York Auckland Cape Town Dar es Salaam Hong Kong Karachi Kuala Lumpur Madrid Melbourne Mexico City Nairobi New Delhi Shanghai Taipei Toronto

With offices in
Argentina Austria Brazil Chile Czech Republic France Greece Guatamala Hungary Italy Japan South Korea Poland Portugal Singapore Switzerland Thailand Turkey Ukraine Vietnam

Oxford is a registered trade mark of Oxford University Press in the UK and in certain other countries

British Library Cataloguing in Publication Data

Data available

ISBN: 978-0-19913631-5

10 9 8 7 6 5 4 3 2 1

Printed by Bell and Bain Ltd., Glasgow

Acknowledgements:
Authors, editors, co-ordinators and contributors: Stephen Pople, Carol Tear, Claire Gordon.

Project managed by Elektra Media Ltd. Typeset by Wearset Ltd.

Paper used in the production of this book is a natural, recyclable product made from wood grown in sustainable forests. The manufacturing process conforms to the environmental regulations of the country of origin.

CONTENTS

Overview

This e-book is a supplementary resource for use alongside your main course book and to help you with your revision. On its own it is not sufficient for full study of your OCR A Level Physics A course. The e-book is a succinct guide to what you need to know and be able to do and will help you to identify any areas that you need to study more thoroughly.

- You can obtain a copy of the specification and past papers to practise. You can also look at the Practical Skills Handbook from the exam boards' website: www.ocr.org.uk.
- Find out when you will be working on the practical skills and the dates of your exams and plan your revision accordingly.
- Begin revising! The self-assessment questions (with answers) on pages 82–84 will help you to check your progress.

Specification structure

		OCR Physics A
AS units	Unit 1	**Mechanics** Motion, forces in action, work and energy *1h written exam* *AS 30% A 15%* *60 marks*
	Unit 2	**Electrons, waves, and photons** Electric current, resistance, DC circuits, waves, quantum physics *1h45m written exam* *AS 50% A 25%* *100 marks*
	Unit 3	**Practical skills in physics 1** *Three externally set tasks, internally marked using set mark scheme* *AS 20% A 10%* *40 marks*
A2 units	Unit 4	**The Newtonian world** Newton's laws and momentum, circular motion and oscillations, thermal physics *1h 15min written exam (synoptic questions)* *A 15%* *60 marks*
	Unit 5	**Fields, particles, and frontiers of physics** Electric and magnetic fields, capacitors and exponential decay, nuclear physics, medical imaging, modelling the universe *2h written exam (synoptic questions)* *A 25%* *100 marks*
	Unit 6	**Practical skills in physics 2** *Three externally set tasks, internally marked using set mark scheme* *A 10%* *40 marks*

What are...

...synoptic questions?

When answering these you will have to apply physics principles or skills in contexts that are likely to be unfamiliar to you. Some questions will require you to show that you understand how different aspects of physics relate to one another or are used to explain different aspects of a particular application. Questions of this type will require you to draw on the knowledge, understanding, and skills developed during your study of the whole course. 20% of the A level marks are allocated to synoptic questions.

Practical assessment

Your practical skills will be assessed at both AS and A2 level.

You are required to carry out three tasks at A2, these are:
A qualitative task (10 marks)
A quantitative task (20 marks)
An evaluative task (10 marks)

OCR will provide a selection of each task and your examination centre (school or college) may give you the opportunity to attempt more than one, so that the best mark can be put forward. The tasks are completed with your teacher supervising. Each practical skills unit is marked by your teacher using instructions from OCR and the marking is moderated (checked) by OCR.

The qualitative and quantitative tasks test skills of observation, recording and reaching valid conclusions. The evaluative task tests your ability to analyse and evaluate the procedures followed and/ or the measurements made. You may be asked to suggest improvements to increase the reliability or accuracy of an experiment. This task will be linked to the quantitative task but you will not have to take any more data. (Additional data may be provided.)

Carrying out practical work

Record
- all your measurements
- any problems you have met
- details of your procedures
- any decisions you have made about apparatus or procedures including those considered and discarded
- relevant things you have read or thoughts you have about the problem.

Define the problem
Write down the aim of your experiment or investigation. Note the variables in the experiment. Define those that you will keep constant and those that will vary.

Suggest a hypothesis
You should be able to suggest the expected outcome of the investigation on the basis of your knowledge and understanding of science. Try to make this as quantitative as you can, justifying your suggestion with equations wherever possible.

Do rough trials
Before commencing an investigation in detail do some rough tests to help you decide on
- suitable apparatus
- suitable procedures
- the range and intervals at which you will take measurements
- consider carefully how you will conduct the experiment in a way that will ensure safety to persons and to equipment.

Remember to consider alternative apparatus and procedures and justify your final decision.

Carry out the experiment
Remember all the skills you have learnt during your course:
- note all readings that you make
- consider carefully the range and intervals at which you make your observations
- take repeats and average whenever possible
- use instruments that provide suitably accurate data
- consider the accuracy to which it is reasonable to quote your observations (how many significant figures are reasonable)
- analyse data as you go along so that you can modify the approach or check doubtful data.

Presentation of data
Tabulate all your observations, remembering to
- include the quantity, any prefix, and the unit for the quantity at the head of each column
- include any derived quantities that are suggested by your hypothesis
- quote measurements and derived data to an accuracy/significant figures consistent with your measuring instruments and techniques. Remember to work out an appropriate unit.
- make sure figures are not ambiguous.

Graph drawing
Remember to
- choose a suitable scale that uses the graph paper fully.
- label your axes with quantity and unit
- use a scale that is easy to use and fills the graph paper effectively
- plot points clearly with a cross using a sharp pencil (you may wish to include 'error bars')
- draw the best line through your plotted points so that the points are scattered evenly about the line or curve
- consider whether the gradient and area under your graph have significance.

Analysing data
This may include
- the calculation of a result
- drawing of a graph
- statistical analysis of data
- analysis of uncertainties in the original readings, derived quantities, and results.

Remember to
- use a large gradient triangle in graph analysis to improve accuracy
- set out your working so that it can be followed easily
- ensure that any quantitative result is quoted to an accuracy that is consisted with your data and analysis methods
- include a unit for any result you obtain.

Evaluation of the investigation
The evaluation should include the following points:
- draw conclusions from the experiment
- identify any systematic errors in the experiment
- comment on your analysis of the uncertainties in the investigation
- review the strengths and weaknesses in the way the experiment was conducted
- suggest alternative approaches that might have improved the experiment in the light of experience.

Use of information technology (IT)
You may have used data capture techniques when making measurements or used IT in your analysis of data. In your analysis you should consider how well this has performed. You might include answers to the following questions.
- What advantages were gained by the use of IT?
- Did the data capture equipment perform better than you could have achieved by a non-IT approach?
- How well has the data analysis software performed in representing your data graphically, for example?

Your Laboratory Notebook
If you write a good report, it should be possible for the reader to repeat what you have done should they wish to check your work.

Use subheadings
These help break up the report and make it more readable. As a guide, the subheadings could be the main sections of the investigation: aims, diagram of apparatus, procedure, etc.

Answering the question

This section contains some examples of types of questions with model answers showing how the marks are obtained. You may like to try the questions and then compare your answers with the model answers given.

MARKS FOR QUALITY OF WRITTEN COMMUNICATION

Quality of written communication is assessed in all units and credit may be restricted if communication is unclear.

You should:

- Make sure that the text is legible and that your spelling, punctuation and grammar are accurate so that the meaning is clear.
- Write your answers in a style that is appropriate for the purpose of answering the question and explaining complex subject matter.
- Organise the information in your answer so that it is clear, and use the proper scientific vocabulary.

ALWAYS SHOW YOUR WORKING

It is wise always to show your working. If you make a mistake in processing the data you could still gain the earlier marks for the method you use.

---

2 marks if your answer
- uses scientific terms correctly
- is written fluently and/or is well argued
- contains only a few spelling or grammatical errors.

An answer that is scientifically inaccurate, is disjointed, and contains many spelling and grammatical errors loses both these marks.

The message is:
do not let your communication skills let you down.

---

1 mark if your answer
- generally uses scientific terms correctly
- generally makes sense but lacks coherence
- contains poor spelling and grammar.

Question 1

Calculation question

The supply in the following circuit has an e.m.f. of 12.0 V and negligible internal resistance.

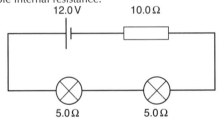

12.0 V   10.0 Ω

5.0 Ω   5.0 Ω

(a) Calculate
  (i) the current through each lamp;   *(2 marks)*
  (ii) the power dissipated in each lamp;   *(2 marks)*
  (iii) the potential difference across the 10.0 Ω resistor.
                                             (1 mark)

(b) A student wants to produce the same potential difference across the 10.0 Ω resistor using two similar resistors in parallel.
  (i) Sketch the circuit the student uses.   *(1 mark)*
  (ii) Determine the value of each of the series resistors used. Show your reasoning.   *(3 marks)*

Answer

(a) (i) Current in circuit = e.m.f./total resistance   (✓)
                  =12.0/20.0
    Current in circuit = 0.60 A   (✓)
  (ii) Power   = I^2R (✓)
               = $0.60^2 \times 5.0$
    Power   = 1.8 W   (✓)
  (iii) p.d. = IR = 0.60×10.0 = 6.0 V   (✓)

(b) (i)

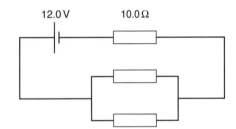

12.0 V   10.0 Ω

Correct circuit as above.   (✓)
(ii) Parallel combination must be 10.0 Ω   (✓)
Two similar parallel resistors have total resistance equal to half that of one resistor.   (✓)
(or $\frac{1}{10} = \frac{1}{R} + \frac{1}{R}$)
Each resistor = 20 Ω   (✓)

Question 2

Experiment description

(a) Sketch the apparatus you would use to show different modes of stationary waves in a stretched string or wire. (2 marks)

(b) Explain what is meant by 'the fundamental mode' (2 marks)

(c) Describe what happens as the frequency of the vibration of the string or wire is increased from zero. (6 marks)

Answer

(a)

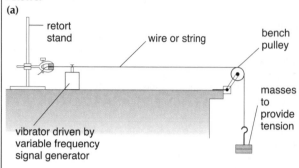

Some means of changing the frequency (✓)

A wire or string in tension (✓)

(b) The fundamental mode is the simplest stationary wave that can be set up. It has the longest wavelength and the lowest frequency. (✓) If the length of the string is l the wavelength of the fundamental mode is $2l$. (✓)

(c) As the frequency is increased to a value f which corresponds to the fundamental, a stationary wave is set up with a node at each end of the string, (✓) and one antinode in the centre. (✓) When the frequency is increased to $2f$ the string will vibrate at the 2nd harmonic (✓) and there is a node in the centre as well as at the ends. (✓) The wavelength is now equal to the length of the string. (✓) When the frequency is increased to $3f$ the string vibrates at the 3rd harmonic and this pattern continues as the frequency is increased to higher harmonics (✓).

Question 3

'Show that' question

A length of wire has diameter 0.5 mm and length 50 cm. The resistance is 2.8 Ω. Show that the resistivity of the wire is about 1×10^{-6} Ωm.

Answer

$\rho = RA/l$ (✓) $A = \pi (0.25 \times 10^{-3})^2$ (✓)

$\rho = 2.8 \times 1.96 \times 10^{-7}/0.5$ Ωm

$\rho = 1.1 \times 10^{-6}$ Ωm (✓)

Measurements, uncertainties and graphs

Significant figures

Writing the value of distance $d = 7$ m does not mean the same as writing $d = 7.0$ m.

$d = 7$ m has 1 significant figure, which implies it could be any value between 6.5 m and 7.4 m.

$d = 7.0$ m has 2 significant figures, which implies it could be any value between 6.95 m and 7.04 m.

Examples

1002 kg has 4 significant figures
0.200 g has 3 significant figures
3.07 ml has 3 significant figures
0.012 g has 2 significant figures

Scientific notation

The average distance from the Earth to the Sun is 150 000 000 km.

There are two problems with quoting a measurement in the above form:

- the inconvenience of writing so many noughts
- uncertainty about which figures are important (i.e. How approximate is the value? How many of the figures are significant?).

These problems are overcome if the distance is written in the form 1.50×10^8 km.

'1.50×10^8' tells you that there are three significant figures – 1, 5, and 0. The last of these is the least significant and, therefore, the most uncertain. The only function of the other zeros in 150 000 000 is to show how big the number is. If the distance were known less accurately, to two significant figures, then it would be written as 1.5×10^8 km.

Numbers written using powers of 10 are in **scientific notation** or **standard form**. This is also used for small numbers. For example, 0.002 can be written as 2×10^{-3}.

Calculations and significant figures

Example

If the distance travelled d is 7.0 m (as above) and the time taken is $t = 2.1$ s (which implies a value in the range $t = 2.05$ s–2.14 s)

The speed $v = \dfrac{d}{t} = \dfrac{7.0\,\text{m}}{2.1\,\text{s}} = 3.3$ ms^{-1} (2sf)

The calculator reads 3.33333… but quoting any more figures would be meaningless because the values of distance and time are not known that accurately. (If you use the range of values for d and t to calculate the biggest and smallest possible values for v, your calculator will show 3.248… and 3.434…).

As a general guide, if the result is to be used in further calculations, it is best to leave any rounding up or down until the end.

Uncertainty

When making any measurement, there is always some **uncertainty** in the reading. As a result, the measured value may differ from the true value. In science, an uncertainty is sometimes called an **error**. However, it is important to remember that it is *not* the same thing as a mistake.

In experiments, there are two types of uncertainty.

Systematic uncertainties These occur because of some inaccuracy in the measuring system or in how it is being used. For example, a timer might run slow, or the zero on an ammeter might not be set correctly.

There are techniques for eliminating some systematic uncertainties. However, this spread will concentrate on dealing with uncertainties of the random kind.

Random uncertainties These can occur because there is a limit to the sensitivity of the measuring instrument or to how accurately you can read it. For example, the following readings might be obtained if the same current was measured repeatedly using one ammeter:

2.4   2.5   2.4   2.6   2.5   2.6   2.6   2.5

Because of the uncertainty, there is variation in the last figure. To arrive at a single value for the current, you could find the mean of the above readings, and then include an estimation of the uncertainty:

current = 2.5 ± 0.1

mean   uncertainty

Writing '2.5 ± 0.1' indicates that the value could lie anywhere between 2.4 and 2.6.

Note:
- On a calculator, the mean of the above readings works out at 2.5125. However, as each reading was made to only two significant figures, the mean should also be given to only two significant figures i.e. 2.5.
- Each of the above readings may also include a systematic uncertainty.

Uncertainty as a percentage

Sometimes, it is useful to give an uncertainty as a percentage. For example, in the current measurement above, the uncertainty (0.1) is 4% of the mean value (2.5), as the following calculation shows:

percentage uncertainty $= \dfrac{0.1}{2.5} \times 100 = 4$

So the current reading could be written as 2.5 ± 4%.

Combining uncertainties

Sums and differences Say you have to *add* two length readings, *A* and *B*, to find a total, *C*. If $A = 3.0 \pm 0.1$ and $B = 2.0 \pm 0.1$, then the minimum possible value of *C* is 4.8 and the maximum is 5.2. So $C = 5.0 \pm 0.2$.

Now say you have to subtract *B* from *A*. This time, the minimum possible value of *C* is 0.8 and the maximum is 1.2 . So $C = 1.0 \pm 0.2$, and the uncertainty is the same as before.

If $C = A + B$ or $C = A - B$, then

uncertainty	=	uncertainty	+	uncertainty
in *C*		in *A*		in *B*

The same principle applies when several quantities are added or subtracted: $C = A + B - F - G$, for example.

Calculated results

Say you have to calculate a resistance from the following readings:

voltage = 3.3 V   (uncertainty ± 0.1 V, or ± 3%)
current = 2.5 A   (uncertainty ± 0.1 A, or ± 4%)

Dividing the voltage by the current on a calculator gives a resistance of 1.32 Ω. However, as the combined uncertainty is ±7%, or ± 0.1 Ω, the calculated value of the resistance should be written as 1.3 Ω (±7%) or 1.3 Ω (±0.1 Ω).

Showing uncertainties on graphs

In an experiment, a wire is kept at a constant temperature. You apply different voltages across the wire and measure the current through it each time. Then you use the readings to plot a graph of current against voltage.

The general direction of the points suggests that the graph is a straight line. However, before reaching this conclusion, you must be sure that the points' scatter is due to random uncertainty in the current readings. To check this, you could estimate the uncertainty and show this on the graph using short, vertical lines called uncertainty bars. The ends of each bar represent the likely maximum and minimum value for that reading. In the example below, the **uncertainty bars** show that, despite the points' scatter, it is reasonable to draw a straight line through the origin.

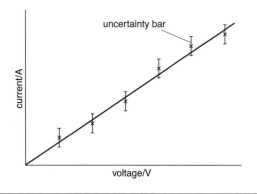

Products and quotients If $C = A \times B$ or $C = A/B$, then

% uncertainty	=	% uncertainty	+	% uncertainty
in *C*		in *A*		in *B*

For example, say you measure a current *I*, a voltage *V*, and calculate a resistance *R* using the equation $R = V/I$. If there is a 3% uncertainty in *V* and a 4% uncertainty in *I*, then there is a 7% uncertainty in your calculated value of *R*.

Note:
- The above equation is only an approximation – and a poor one for uncertainties greater than about 10%.
- To check that the equation works, try calculating the maximum and minimum values of *C* if, say, *A* is 100 ± 3 and *B* is 100 ± 4. You should find that $A \times B$ is 10 000 ± approximately 700 (i.e. 7%).
- The principle of adding % uncertainties can be applied to more complex equations: $C = A^2B/FG$, for example. As $A^2 = A \times A$, the % uncertainty in A^2 is twice that in *A*.

Choosing a graph

The general equation for a straight-line graph is

$$y = mx + c$$

In this equation, *m* and *c* are **constants**, as shown below. *y* and 3 are **variables** because they can take different values. x is the **independent variable**. *y* is the **dependent variable**: its value depends on the value of x.

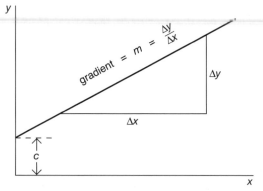

In experimental work, straight-line graphs are especially useful because the values of constants can be found from them.

Labelling graph axes Strictly speaking, the scales on the graph's axes are pure, unitless numbers and not voltages or currents. Take a typical reading:

voltage = 10 V

This can be treated as an equation and rearranged to give:

voltage/V = 10

That is why the graph axes are labelled 'voltage/V' and 'current/A'. The values of these are pure numbers.

Reading a micrometer

The length of a small object can be measured using a micrometer screw gauge. You take the reading on the gauge like this:

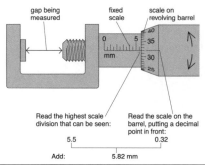

Reading a vernier

Some measuring instruments have a vernier scale on them for measuring small distances (or angles). You take the reading like this:

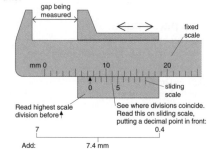

How Science Works

There are 12 aspects to 'How Science Works' that are included in all the AS and A level specifications. Some of the specifications have incorporated these ideas into different modules, and some have rewritten them so that the wording is different.
The 12 aspects are listed here with the original wording, together with some guidance on interpretation, and the type of questions that could be asked.

Use theories, models, and ideas to develop and modify scientific explanations

There are many historical examples of scientists making some observations and then using creative thinking and imagination to interpret the data and develop an explanation. The first step is to come up with an **idea** – an initial thought about the reasons for the observations. This is then extended and worked into a model, maybe combining several ideas.

One definition of a **model** is:

A representation of a system that allows for investigation of the properties of the system and, in some cases, prediction of future outcomes.

The model is then tested and, if it works, can be set out as a **theory**.

One definition of a scientific theory is:

A set of statements or principles that explain observations, especially a set that has been repeatedly tested or is widely accepted and can be used to make predictions about natural phenomena.

You may be asked to give an example. Here are some:

Galileo Galilei timed objects rolling down an inclined plane and concluded that falling objects accelerate. Isaac Newton suggested a model for gravity and showed that freely-falling objects have the same acceleration. He developed the theory of gravity.

Scientists often use microscopic models, such as that of particle behaviour, to explain macroscopic behaviour. For example, the kinetic theory of gases explains the gas laws.

Other examples of using models to develop theories include:

- Newton's laws
- the Rutherford model of the atom
- Hubble's law.

At this stage in your study of science you are unlikely to be thinking up new theories, but you may be doing experiments to see if your observations fit with a model. For example, plotting the square of the period, T, of a simple pendulum against its length, l, to see if this gives a straight line through the origin. If it does, it shows that $T^2 \propto l$, which confirms that the oscillations are an example of simple harmonic motion.

Use knowledge and understanding to pose scientific questions, define scientific problems, present scientific arguments and scientific ideas

As part of your course you use scientific theories to answer scientific questions or address scientific problems. In addition, you are expected to identify scientific questions or problems (within a given context). You may be presented with a hypothesis (an untested theory based on observations) or be asked to suggest one. The hypothesis needs to be tested by experiment, and if a reliable experiment does not support a hypothesis it must be changed.

When presenting arguments and ideas you should be able to

distinguish between questions that science can address, and those that science cannot address. For example, whether a view is beautiful is not a question science can answer.

A historical example is the photoelectric effect.

The question was 'For a metal that shows the photoelectric effect, why is there a threshold frequency? (Why does high intensity red light cause no emission of photoelectrons, but low intensity UV radiation does cause emission?)' Albert Einstein suggested a hypothesis based on Max Planck's ideas: that the radiation was quantized and arrived in packets called photons, with energy $E = hf$.

Use appropriate methodology, including ICT, to answer scientific questions and solve scientific problems

This includes how you conduct experimental work for example:

- planning, or following a plan, of an investigation
- identifying the dependent and independent and control variables
- selecting appropriate apparatus and methods (including ICT) to carry out reliable experiments
- choosing instruments with appropriate sensitivity and precision (see opposite)
- justifying the methods used during experiments (including the use of ICT) to collect valid and reliable data and produce scientific theories
- using ICT (spreadsheets, for example) to develop scientific models or plot graphs, and dataloggers to monitor physical changes.

Sensitivity and precision
Take the example of measuring mass. The more *sensitive* a balance is, the smaller variation in mass the balance can detect and measure. A mass smaller than the sensitivity of a balance is not detectable using the balance.

If the mass of an object is measured many times, the *precision* is indicated by the spread of the results. If the measurements are all very close, the precision of the instrument is greater.

Accuracy and precision
A measurment with great precision is not the same as one with great accuracy, as illustrated by the diagram:

Not accurate or precise

Accurate but not precise

Precise but not accurate

Accurate and precise

Carry out experimental and investigative activities, including appropriate risk management, in a range of contexts

You should be able to show that you can:

- follow experimental procedure in a sensible order
- use appropriate apparatus and methods to make accurate and reliable measurements (see above)
- identify and minimize significant sources of experimental error
- identify and take account of risks in carrying out practical work
- produce a risk assessment before carrying out a range of practical work.

An example would be to recognize that you should use forceps when handling radioactive sources and replace them in the lead-lined protective, locked case when you have finished.

Analyse and interpret data to provide evidence, recognizing correlations and causal relationships

This refers to the methods you use in your experimental work, and what you do with the data you collect.

You will be expected to:

- record data in tables, and sometimes use equations to calculate values that you add to the table (for example, if you measure the extention of a wire you might add strain, ε to your table
- plot and use graphs to establish or verify relationships between variables
- calculate the gradient and find the intercepts of straight-line graphs
- analyse data, including graphs, to identify patterns and relationships (correlation and cause, for example).

Correlation and cause
Analyse graphs of datasets that show different correlations.

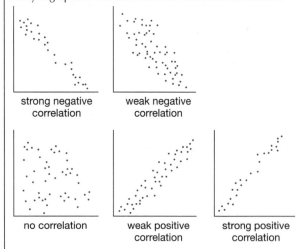

strong negative correlation   weak negative correlation

no correlation   weak positive correlation   strong positive correlation

Remember that a correlation does not necessarily show that one thing causes the other. (Children with larger feet are better at spelling but this is because older children have larger feet and are better at spelling then younger children, not because larger feet *cause* better spelling).

Some of the more complex analysis included here is the use of log graphs to analyse nuclear decay or the discharge of a capacitor.

Evaluate methodology, evidence and data, and resolve conflicting evidence

You should be able to:

- recognize, and distinguish between, systematic and random errors
- estimate the errors in measurements
- use data, graphs, and other experimental evidence to draw conclusions
- use the (estimated) most significant error to assess the reliability of your conclusion
- evaluate the validity of conclusions in the light of the experimental methods used
- recognize conflicting evidence.

These skills can be applied to any experiments, such as using an airtrack to do experiments involving the acceleration and the change of momentum in collisions of airtrack gliders.

Appreciate the tentative nature of scientific knowledge

In everyday speech, 'tentative' usually means hesitant or unsure, but it also means 'not fully worked out', or 'a work in progress'. This is the nature of scientific knowledge.

Once scientists find a theory that works well, they accept and use it for as long as it works, but they recognize and accept that if new observations in the future conflict with the theory, and a better explanation is offered, then the accepted theory will change. This is unlike some other types of knowledge or belief.

A good example is our model of the atom, which has evolved over many years with contributions from many scientists. During experiments, when events occur that cannot be explained using current scientific theories (such as the alpha particle back-scattering observed by Hans Geiger and Ernest Marsden), scientists modify existing theories, or produce new theories to explain their findings (such as Ernest Rutherford's nuclear model of the atom).

Another example is the way in which Newton's laws of motion, which work perfectly well for everyday speeds, do not correctly describe behaviour at speeds approaching light speed. The modification of our theory of motion to take into account Einstein's theory of relativity was necessary.

Einstein's theory was originally theoretical, but it predicted distortion of starlight during an eclipse, which was observed in 1919. In 1971 accurate atomic clocks were available that were sensitive enough to detect the difference in the time taken by clocks flown in different directions around the world for 3 days. When you use a satellite navigation system, the processing takes into account relativity in determining your position.

Other examples of the development of theories include:

- theories of motion – Galileo and Aristotle
- the photoelectric effect
- the binding energy of the nucleus
- Olber's paradox and other cosmological questions (e.g. What is dark matter?)
- the search for the Higgs boson – a theoretical particle the existence of which may explain why particles have mass
- the prediction of antiparticles in 1930 before the discovery of the positron in 1932.

Communicate information and ideas in appropriate ways using appropriate terminology

Using the correct scientific terminology avoids confusion. You should be able to write explanations using correct scientific terms, and support your arguments with equations, diagrams and clear sketch graphs. This applies to exam papers, practical work, and investigations.

Consider applications and implications of science and appreciate their associated benefits and risks

You may be asked to discuss the risk associated with an activity from almost any physics topic, in terms of the actual level of the risk and its potential consequences. People's perception of risk depends on factors such as how familiar the risk is. For example, people overestimate the risk of an aircraft crash, and underestimate the risk of a car crash.

There are some physics topics with more obvious benefits and risks than others.

Fossil fuels, electricity generation, and global warming

Since the Industrial Revolution, the application of physics has included many activities that involve burning fossil fuels, which release carbon dioxide. Most scientists now think that global warming is at least partly due to human activities. You should be aware of the impact this has had on the environment and how scientists are using their current findings to inform decision-makers of the consequence of global warming and advise them how to minimize its effects.

Medical treatment and imaging

Some medical imaging techniques, for example X-rays, positron emission tomography (PET) scans, and the use of radioactive tracers, involve risks to staff and patients from ionizing radiation. Magnetic resonance imaging (MRI) scanners use very large magnetic fields from superconducting magnets, and there is a risk because any metal will be strongly attracted to these magnets. The benefits to the patient may be a diagnosis without surgery.

Cancer treatment using radioisotopes also involves a risk from ionizing radiation, but the benefit of a cure outweighs the risk.

Using nuclear reactors to provide electrical power

The benefit of nuclear reactors is a large amount of electrical energy for a small amount of nuclear fuel. The risks are associated with ionizing radiation. There is concern about contamination of the environment during operation, and also from the nuclear waste, which has long half-lives.

Ultraviolet radiation

Radiation reaching us from the Sun includes UVA, which has an aging effect, and UVB, which can damage the cornea of the eye and cause skin cancer. Sunlight is needed by the body to produce vitamin D, which is required for bone growth in children and has a role in reducing the risk of other cancers. Ultraviolet radiation with shorter wavelengths, UVC, is more dangerous. Sunscreen filters out the ultraviolet radiation so that it does not reach the skin.

Other topics include:
- car safety, including the global positioning system (GPS), which has benefits we use everyday in satellite navigation systems, but the system can also be used to target air strikes accurately
- material properties
- geostationary satellites
- resonance
- mass spectrometry
- use of radioactive isotopes
- capacitors used for flash photos, lasers for fusion research, and back-up power supplies for computers.

Consider ethical issues in the treatment of humans, other organisms, and the environment

You should be able to identify ethical issues arising from the application of science as it impacts on humans and the environment, and discuss scientific solutions from a range of ethical viewpoints.

Scientific research is funded by society, either through public funding or through private companies that obtain their income from commercial activities. Scientists have a duty to consider ethical issues associated with their findings. They set up groups to decide what should be permitted, and also contribute to groups set up by society to make decisions about what should be permitted.

Individual scientists have ethical codes that are often based on humanistic, moral, and religious beliefs.

Science has provided solutions to problems, but it is up to society as a whole (including scientists) to judge whether the solution is acceptable in view of the moral issues that result. Issues such as effects on the planet, and the economic and physical well-being of the living things on it, should be considered. Secure transmission of data is important if people are to be confident that personal data cannot be intercepted in transmission.

When a country is at war there may be difficult decisions for scientists to make. In the Second World War, scientists on both sides were in a race to build the first atom bomb.

Music can now be stored and reproduced to a high standard, but as a result infringement of copyright has become simple. Steps have been taken to make it more difficult to download music illegally – an example of science and technology providing solutions to the problems it creates.

Appreciate the role of the scientific community in validating new knowledge and ensuring integrity

It is important that new data, and new interpretations of data, should be critically evaluated. This is true whether they support established scientific theories or propose new theories.

Scientists communicate their findings to other scientists through journals and conferences. By sharing the findings of their research, scientists provide the scientific community with opportunities to replicate and further test their work. This can result in either confirming new explanations or refuting them.

Peer review

Some scientific journals state in the journal (and on their websites) that they are peer reviewed.

This means that the papers submitted by scientists for publication are sent for peer review before being accepted for publication. A peer is 'a person who is of equal standing with another in a group.' In this case, it is another scientist, or scientists, working in the same, or similar, field of research. Other scientists know that everything in the journal has been considered by another qualified, independent, scientist.

Note that some scientific magazines are not peer reviewed, but may use peer-reviewed articles as a source of information, as well as accepting other articles. A scientist reporting new research would usually publish in a peer-reviewed journal first.

Funding

The interests of the organizations that fund scientific research can influence the direction of that research. In some cases, the validity of the resulting claims may also be influenced.

The UK Government's leading funding agency for research and training in engineering and the physical sciences is the Engineering and Physical Science Research Council (EPSRC). Scientists submit proposals detailing the research they want to do and the equipment and staff they need.

The proposals are peer reviewed, and then all the proposals and the reviewer's comments are considered by a committee of scientists.

Almost all scientists work with the common aim of progressing scientific knowledge and understanding in a valid way and believe that accurate reporting of findings should take precedence over recognition of success of an individual. However, a disadvantage of the system could be that less ethical reviewers have the opportunity to benefit from other scientists' ideas and results, and to prevent publication and funding of their work. To prevent this, the reviewers' comments are sent to the author/researcher for comment. The system could work to exclude scientists with unusual research ideas or theories.

Cold fusion

In 1989, Martin Fleischmann and Stanley Pons reported a nuclear fusion reaction that took place during the electrolysis of heavy water (water containing the hydrogen isotope deuterium, ^2_1H) using palladium electrodes. Heat was produced, which they claimed was due to nuclear fusion of deuterium. There was great interest all over the world because of the possibility of a cheap and abundant source of power. (It was called 'cold fusion' because nuclear fusion research concentrates on producing the high temperatures necessary to bring nuclei close enough to fuse.) The discovery was rejected after other scientists were unable to reproduce the discovery.

Examples of different parts of the scientific community working together include the following:

- The experimental discovery of electron diffraction confirmed the dual nature of matter particles, first put forward by de Broglie as a hypothesis several years earlier.
- In the search for a unifying theory, scientists make new discoveries based on theoretical predications, and continue to work to confirm the discovery of others, such as the Higgs boson.

Appreciate the ways in which society uses science to inform decision-making

Science influences decisions on an individual, local, national, and international level.

Scientific findings lead to new technologies, which enable advances to be made that have potential benefit for humans. However, these have to be balanced against the risks.

In practice, the scientific evidence available to decision-makers may be incomplete – scientific evidence should be considered as a whole. Decision-makers, who include government-appointed science advisers, are influenced by many things. These include their prior beliefs, their vested interests, special interest groups, public opinion, and the media, as well as by expert scientific evidence. The media and pressure groups often select parts of scientific evidence that support a particular viewpoint. This can influence public opinion, which in turn may influence decision-makers. Consequently, decision-makers may make socially and politically acceptable decisions based on incomplete evidence. The following are examples of this.

Electric cars may replace petrol vehicles if batteries are developed that give a greater range than those at present.

Until then, car buyers are unlikely to be persuaded to buy electric cars.

Satellite tracking for purposes such as road pricing may be implemented without adequate trials because of pressure group influence.

The improved communication that digital electronics bring to society means that people can find out more easily what is happening and give their views. (For example, there are many petitions on the Downing Street website that you can sign online.) The range of information made available to decision-makers in industry, services, and government has increased now that information can be processed and presented using computers.

The expense of space travel is one area in which people have strong views. Some people regard it as a waste of money that could be used for building hospitals, for example. Others see the interest generated from space travel as beneficial and the technological spin-offs as worthwhile. Space travel contributes to global warming, but also to a greater understanding of climate on Earth and other planets.

Nuclear power, and the safe disposal of nuclear waste.

Unit G484
The Newtonian world

Module 1: Newton's laws and momentum

4.1.1 Newton's laws of motion

Newton's first law

A force is needed to change the motion of an object. This idea is summed up by *Newton's first law of motion*:

> If there is no resultant force acting,
> * a stationary object will stay at rest
> * a moving object will maintain a constant velocity (a steady speed in a straight line).

From Newton's first law, it follows that if an object is at rest or moving at constant velocity, then the forces on it must be balanced, as in the examples above.

The more mass an object has, the more it resists any change in motion (because more force is needed for any given acceleration). Newton called this resistance to change in motion *inertia*.

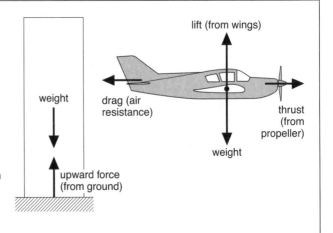

Newton's third law

A single force cannot exist by itself. Forces are always pushes or pulls between *two* objects, so they always occur in pairs. One force acts on one object; its equal but opposite partner acts on the other. This idea is summed up by *Newton's third law of motion*:

> If A is exerting a force on B, then B is exerting an equal but opposite force on A.

Examples of action–reaction pairs are given below.

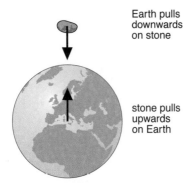

Note:
* It does not matter which force you call the action and which the reaction. One cannot exist without the other.
* The action and reaction do not cancel each other out because they are acting on *different* objects.

Momentum and Newton's second law

The product of an object's mass m and velocity v is called its linear *momentum, p*:

> momentum = mass × velocity

$$p = m v$$

Momentum is measured in kg m s^{-1}. It is a vector.

According to *Newton's second law of motion*:

> The rate of change of momentum of an object is proportional to the resultant (net) force acting on it.

This can be written in the following form:

$$\text{resultant force} \propto \frac{\text{change in momentum}}{\text{time taken}}$$

If the change in momentum is Δp and the time taken is Δt, the resultant force F is given by:

$$F = k \frac{\Delta p}{\Delta t}$$

where k is a constant value.

With the unit of force defined in a suitable way (as in SI), the value of $k = 1$ so:

$$F = \frac{\Delta p}{\Delta t} \qquad (1)$$

Example

The tennis ball in the diagram has change in momentum = final momentum – initial momentum:

$$\Delta p = mv - mu$$
$$F = \frac{mv - mu}{\Delta t} \qquad (2)$$

initial velocity: u →

final velocity: v →

m → F

Aircraft propulsion

To move forward, an aircraft pushes a mass of gas backwards so that, by Newton's third law, there is an equal forward force on the aircraft. Here are two ways of producing a backward flow of gas:

Jet engine Air is drawn in at the front by a large fan, and pushed out at the back. Exhaust gases are also ejected, at a higher speed.

Propeller This is driven by the shaft of a jet engine or piston engine. Its blades are angled so that air is pushed backwards as it rotates.

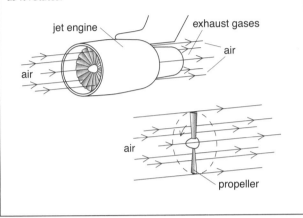

Momentum problem

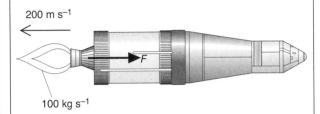

Example A rocket engine ejects 100 kg of exhaust gas per second at a velocity (relative to the rocket) of 200 m s^{-1}. What is the forward thrust (force) on the rocket?

By Newton's third law, the forward force on the rocket is equal to the backward force pushing out the exhaust gas. By Newton's second law, this force F is equal to the momentum gained per second by the gas, so it can be calculated using the equation $F = \dfrac{\Delta p}{\Delta t}$ with the following values:

$$m = 100 \text{ kg} \quad \Delta t = 1.0 \text{ s} \quad u = 0 \quad v = 200 \text{ m s}^{-1}$$

So $F = \dfrac{\Delta p}{\Delta t} = \dfrac{mv - mu}{\Delta t}$

$$F = \frac{(100 \times 200) - (100 \times 0)}{1.0} = 20\ 000 \text{ N}$$

Newton's second law – for constant mass

When a resultant force acts on an object, its velocity changes. Newton's second law applies to situations in which the mass and velocity both change – for example, a rocket using fuel so that the mass reduces as the rocket accelerates.

In the special case of the mass staying constant, the equation can be written as

$$F = \frac{m(v - u)}{t}$$

But acceleration $a = \dfrac{(v - u)}{t}$ (3)

So, when mass is constant, $F = ma$

This equation comes from Newton's second law, but it is not correct to say it is Newton's second law – because the equation can only be used when the mass is constant.

Note:
- According to Einstein, mass increases with velocity (though insignificantly for velocities much below that of light). This means that $F = ma$ is really only an approximation, though an acceptable one for most practical purposes.

- When using equations (1), (2) and (3), remember that F is the net force acting on the body. For example, below, the resultant force is 26.0 – 20.0 = 6.0 N upwards. The upward acceleration a can be worked out as follows:

$$a = \frac{F}{m} = \frac{6.0}{2.0} = 3.0 \text{ m s}^{-2}$$

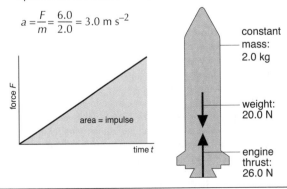

Impulse

Equation (1) can also be rewritten $F\Delta t = \Delta p$

In words   force x time taken = change in momentum

The quantity $F\Delta t$ is called an *impulse*.

A given impulse always produces the same change in momentum, irrespective of the mass. For example, if a resultant force of 6.0 N acts for 2.0 s, the impulse delivered is $6.0 \times 2.0 = 12.0$ N s.

This will produce a momentum change of 12 kg m s^{-1}.

So a 4.0 kg mass will gain 3.0 m s^{-1} of velocity

or a 2.0 kg mass will gain 6.0 m s^{-1} of velocity, and so on.

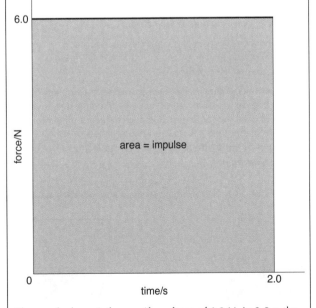

The graph above is for a uniform force of 6.0 N. In 2.0 s, the impulse delivered is 12 N s. Numerically, this is equal to the area of the graph between the 0 and 2.0 s points.

4.1.2 Collisions

Conservation of momentum

Trolleys A and B below are initially at rest. When a spring between them is released, they are pushed apart.

By Newton's third law, the force exerted by A on B is equal (but opposite) to the force exerted by B on A. These equal forces also act for the same time, so they deliver equal (but opposite) impulses. As a result, A gains the same momentum to the left as B gains to the right.

Before separation

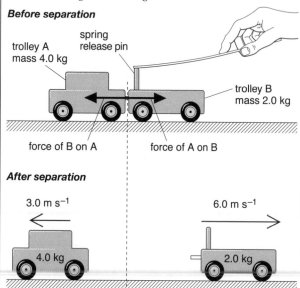

After separation

Momentum is a vector, so its direction can be indicated using + or −. If vectors to the right are taken as +,

> before the trolleys separate
>   total momentum = 0
>
> after the trolleys separate
>   momentum of A = $4.0 \times (-3.0) = -12$ kg m s^{-1}
>   momentum of B = $2.0 \times (+6.0) = +12$ kg m s^{-1}
>   so total momentum = 0 kg m s^{-1}

Together, trolleys A and B make up a **system**. The total momentum of this system is the same (zero) before the trolleys push on each other as it is afterwards. This illustrates the **law of conservation of momentum**:

> When the objects in a system interact, their total momentum remains constant, provided that there is no external force on the system.

Below, the separating trolleys are shown with velocities of v_1 and v_2 instead of actual values. In cases like this, it is always best to choose the same direction as positive for all vectors. It does not matter that A is really moving to the left. If A's velocity is 3.0 m s^{-1} to the left, then $v_1 = -3.0$ m s^{-1}.

After separation

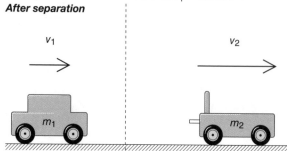

As the total momentum of the trolleys is zero,

$$m_1 v_1 + m_2 v_2 = 0$$

So, if v_2 is positive, v_1 must be negative.

Momentum in collisions

Whenever objects collide, their total momentum is conserved, provided that there is no external force acting.

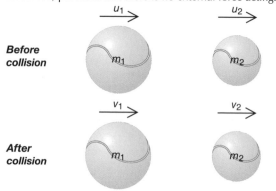

Above, two balls collide and then separate. All vectors have been defined as positive to the right. As the total momentum is the same before and after,

$$m_1 u_1 + m_2 u_2 = m_1 v_1 + m_2 v_2 \qquad (1)$$

Kinetic energy in collisions

Elastic collision An elastic collision is one in which the total kinetic energy of the colliding objects remains constant. In other words, no energy is converted into heat (or other forms). If the above collision is elastic,

$$\tfrac{1}{2}m_1 u_1^2 + \tfrac{1}{2}m_2 u_2^2 = \tfrac{1}{2}m_1 v_1^2 + \tfrac{1}{2}m_2 v_2^2$$

One consequence of the above is that the speed of separation of A and B is the same after the collision as before:

$$u_1 - u_2 = -(v_1 - v_2)$$

Inelastic collision In an inelastic collision, kinetic energy is converted into heat. The total amount of *energy* is conserved, but the total amount of *kinetic energy* is not.

Energy or momentum

Energy and momentum are completely different quantities. They have different units. Energy is calculated from force × distance. It is measured in joules. One joule of kinetic energy is gained when a force of 1 newton acts on an object over a distance of 1 m.

Momentum is calculated from force × time. It is measured in newton seconds (or kg m s^{-1}, which is the same). One newton second of momentum is gained when a force of 1 newton acts on an object for 1 second. Momentum is a vector and has direction as well as magnitude. Energy is a scalar and has magnitude only.

The equation $E_k = \tfrac{1}{2}mv^2$ and $p = mv$, so

$$E_k = \frac{p^2}{2m}$$

When considering collisions remember the following.

The **principle of conservation of momentum** says that the total momentum is conserved if there is no external force acting. This also results from Newton's third law (the force on each of the two objects is equal and opposite) and the fact that the forces must act for the same time on each of the objects. In an **elastic collision** kinetic energy is conserved. In an inelastic collision kinetic energy is not conserved. You must specify that it is what happens to the *kinetic* energy that determines whether a collision is elastic or inelastic, because the **principle of conservation of energy** says that energy is always conserved. In collisions some kinetic energy is usually transferred, for example to heat and sound, but the total energy is unchanged.

Collision problems

Before collision

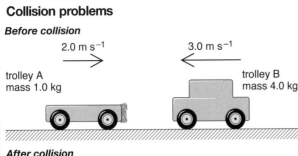

2.0 m s^{-1}

3.0 m s^{-1}

trolley A
mass 1.0 kg

trolley B
mass 4.0 kg

After collision

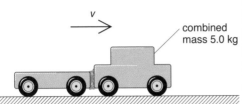

v

combined
mass 5.0 kg

Example 1 *The trolleys above collide and stick together. What is their velocity after the collision? (Assume no friction.)*

All vectors to the right will be taken as positive.
The unknown velocity is v (to the right).
    momentum = mass × velocity

before the collision
momentum of A = $1.0 \times 2.0 = 2.0$ kg m s^{-1}
momentum of B = $4.0 \times (-3.0) = -12$ kg m s^{-1}
∴ total momentum = -10 kg m s^{-1} (2)

After the collision
A and B have a combined mass of 5.0 kg, and a combined velocity of v. So total momentum = $5.0 \times v$.

As the total momentum is the same before and after,

    $5.0v = -10$ which gives $v = -2.0$ m s^{-1}

So the trolleys have a velocity of 2.0 m s^{-1} to the *left*.

Example 2 *When the trolleys collide, how much of their total kinetic energy is lost (converted into other forms)?*

    $E_k = \frac{1}{2}mv^2$

before the collision
E_k of A = $\frac{1}{2} \times 1.0 \times 2.0^2 = 2.0$ J
E_k of B = $\frac{1}{2} \times 4.0 \times (-3.0)^2 = 18$ J
∴ total E_k = 20 J   (3)

after the collision
total E_k = $\frac{1}{2} \times 5.0 \times (-2.0)^2 = 10$ J

Comparing the total E_k before and after, 10 J of E_k is lost.

Example 3 *If the collision had been elastic, what would the velocities of the trolleys have been after separation?*

Let v_1 be the final velocity of A and v_2 be the final velocity of B (both defined as positive to the right).

As both total momentum and total E_k are conserved,

    total momentum after collision   = -10 kg m s^{-1} (from 2)
    total E_k after collision   = 20 J   (from 3)

So $(1.0 \times v_1) + (4.0 \times v_2)$ = -10
And $(\frac{1}{2} \times 1.0 \times v_1^2) + (\frac{1}{2} \times 4.0 \times v_2^2)$ = 20
Solving these equations for v_1 and v_2 gives

    $v_1 = -6.0$ m s^{-1} and $v_2 = -1.0$ m s^{-1}

Note:
- There is an alternative solution which gives the velocities before the collision: 2.0 m s^{-1} and –3.0 m s^{-1}, which is not possible, as it means the trolleys have passed through each other.

Recoiling particles

Before split

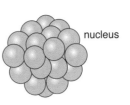

nucleus

After split

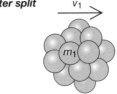

v_1

v_2

m_1

m_2

Above, an atomic nucleus splits into two smaller particles with a loss of nuclear energy. The particles share the energy released (as kinetic energy) and shoot apart. All vectors have been defined as positive to the right.

As the total momentum is conserved, $m_1 v_1 + m_2 v_2 = 0$ (4)

Also E_k of A = $\frac{1}{2}m_1 v_1^2$ (5)

and E_k of B = $\frac{1}{2}m_2 v_2^2$ (6)

From (4), (5), and (6), the following can be obtained:

    $\dfrac{E_k \text{ of A}}{E_k \text{ of B}} = \dfrac{m_2}{m_1}$

This means, for example, that if A has 9 times the mass of B, then B will shoot out with 9 times the E_k of A. In other words, it will have 90% of the available energy. The energy is only shared equally if A and B have the same mass.

Reducing collision damage

When a vehicle comes to a sudden stop, the momentum and kinetic energy of the vehicle and the occupants are reduced to zero very rapidly. There are two ways of looking at this.

To reduce the kinetic energy, work is done on the person over a very small distance, which means the force is very large ($\Delta E_k = W = Fs$).

To reduce the momentum there is an impulse on the person for a very short time, which means the force is very large ($\Delta p = Ft$).

The force on the occupants can be reduced, while still transferring the same amount of kinetic energy, by increasing the distance over which they are brought to a stop.

The force on the occupants can be reduced, while still reducing the momentum to zero, by increasing the time over which the change of momentum occurs.

Safety features, such as air bags, seat belts and crumple zones, make use of these ways of reducing the force.

Other examples are:

- bending your knees when you land from jumping, to increase the distance and time over which your upper body is brought to a stop

- moving your hands with a ball when you catch it to prevent it stinging your hands

- train buffers compressing and bringing the train to a stop

- safety helmets containing padding that compresses and stops the head hitting a hard surface.

Module 2: Circular motion and oscillations

4.2.1 Circular motion

The radian

On the right, the angle θ in **radians** is defined like this:

$$\theta = \frac{s}{r}$$

s/r has no units because
$m \times m^{-1} = 1$. However, when measuring an angle in radians, a unit should be included for clarity: 2 rad, for example.
When $\theta = 1$ $s = r$ so

1 radian is the angle at the centre of a circle that subtends an arc equal to the radius.

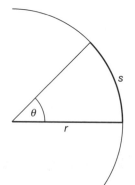

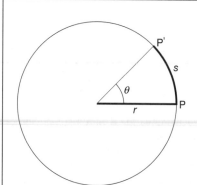

Angular displacement

If point P moves to P′, then the angle θ is called the **angular displacement**. It is measured in **radians**:

$$\theta = \frac{s}{r}$$

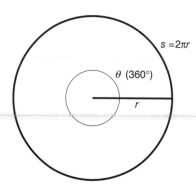

If s is the full circumference of the circle,

$$\theta = \frac{2\pi r}{r} = 2\pi$$

So 2π radians = $360°$

$\therefore$ 1 radian = $\dfrac{360}{2\pi} = 57.3°$

Example

$$60° = \frac{60° \times 2\pi}{360°} = \frac{\pi}{3} \quad \text{radians}$$

Example

$$\frac{\pi}{4} \text{ radians} = \frac{360°}{2\pi} \times \frac{\pi}{4} = 45°$$

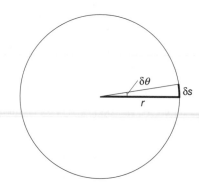

Above, $\delta\theta$ is a very small angle. ($\delta\theta$ counts as one symbol.) δs is so small that it can either be the arc of a circle or the side of a triangle. So

$$\sin \delta\theta = \frac{s}{r} = \delta\theta$$

i.e. for *small* angles $\sin \delta\theta = \delta\theta$.

Rate of rotation

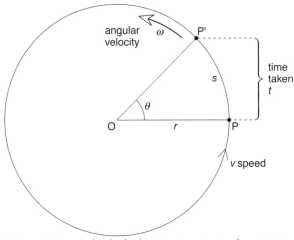

P is a point on a wheel which is turning at a steady rate. In time t, it moves to P'. The rate of rotation can be measured either as an **angular velocity** or as a **frequency**.

Angular velocity ω

$$\text{angular velocity} = \frac{\text{angular displacement}}{\text{time taken}}$$

In symbols $\qquad \omega = \dfrac{\theta}{t}$

For example, if a wheel turns through 10 radians in 2 seconds, then $\omega = 5$ rad s^{-1}. Angular frequency is also measured in rad s^{-1}. It is the magnitude of the vector angular velocity.

Frequency f

$$\text{frequency} = \frac{\text{number of rotations}}{\text{time taken}}$$

Frequency is measured in hertz (Hz). For example, if a wheel completes 12 rotations in 4 seconds, then $f = 3$ Hz.

Period T This is time taken for one rotation. If a wheel makes 3 complete rotations per second ($f = 3$ Hz), then the time taken for one rotation is $\frac{1}{3}$ second. So

$$T = \frac{1}{f}$$

Linking ω, f, and T As there are 2π radians in one full rotation (360°),

$$\omega = 2\pi f$$

For example, a wheel turning at 3 rotations per second ($f = 3$ Hz) has an angular velocity of 6π radians per second.

As $T = 1/f$, it follows from the previous equation that

$$T = \frac{2\pi}{\omega}$$

Velocity in a circle

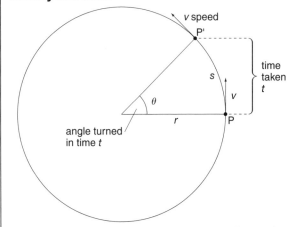

Above, a particle is moving in circle with a steady speed v. (It is not a steady velocity because the direction of the velocity vector is changing.) The particle moves a distance s in time t, so

$$v = \frac{s}{t}$$

The circumference of the circle = $2\pi r$ and this is one revolution of the circle; the distance moved in the period T

So $v = \dfrac{2\pi r}{T}$

Example

A wheel of radius 4.0 cm completes 12 rotations in 4.0 seconds. How fast is a point on the rim moving?

Period $\quad T = (4.0 \div 12)$ s $= 0.333$ s $= 0.33$s (2sf)

$v = \dfrac{2\pi \times 4.0 \text{ cm}}{0.333 \text{ s}} = 75$ cm s^{-1} or 0.75 m s^{-1}

Centripetal acceleration

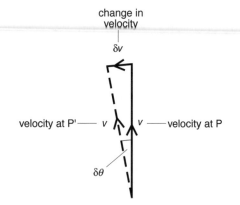

Above, a particle is moving in a circle with a steady speed v. The diagram shows how the velocity vector changes direction as the particle moves from P to P' in time δt.

If θ is in radians, the distance moved is $r\delta\theta = v\delta t$

So time $\delta t = \dfrac{r\delta\theta}{v}$

Below, the velocity vectors from the previous diagram have been used in a triangle of vectors.

The δv vector represents the *change* in velocity because it is the velocity vector which must be *added* to the velocity at P to produce the new velocity (the resultant) at P'. Note that the change in velocity is towards O. In other words, the particle has an *acceleration* towards the centre of the circle. This is called **centripetal acceleration**.

If a is the centripetal acceleration, $\qquad a = \dfrac{\delta v}{\delta t}$ \qquad (1)

But, from the triangle above, $\delta\theta = \dfrac{\delta v}{v}$. So $\delta v = v\delta\theta$

Substituting for δv in equation (1), $\qquad a = \dfrac{v\delta\theta}{\delta t}$

substituting for δt in equation (1) $\qquad a = \dfrac{v\delta\theta}{\left(r\frac{\delta\theta}{v}\right)}$

So

$$a = \frac{v^2}{r}$$

For example, if a particle is moving at a steady speed of 3 m s^{-1} in a circle of radius 2 m, its centripetal acceleration a is found using the middle equation: $a = 3^2/2 = 4.5$ m s^{-2}.

Note:
- When something accelerates, its velocity changes. As velocity is a vector, this can mean a change in *speed* or *direction* (or both). Centripetal acceleration is produced by a change in direction, not speed.

Centripetal force

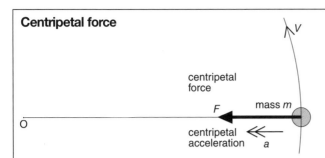

To produce centripetal acceleration, a **centripetal force** is needed. It must act towards the centre of the circle. The centripetal force F, mass m, and centripetal acceleration a are linked by the equation $F = ma$. So, using the equation for a

$$F = \frac{mv^2}{r}$$

Note:

* Centripetal force is *not* produced by circular motion. It is the force *needed* for circular motion. Without it, the object would travel in a straight line.

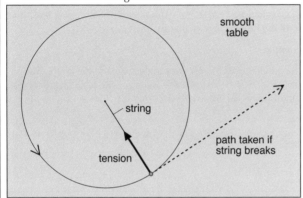

Above, a mass moves in a circle on a smooth table. The tension in the string provides the centripetal force needed. There is no outward 'centrifugal force' on the mass. If the string breaks, the mass travels along a tangent.

Angle of bank An aircraft must bank to turn. This is so that the lift L (from the wings) and the weight mg can produce a resultant to provide the centripetal force F, where

$$F = \frac{mv^2}{r}$$

In the triangle of vectors (below right):

$$L \cos \theta = mg \quad \text{and} \quad L \sin \theta = \frac{mv^2}{r}$$

Dividing the second equation by the first gives $\tan \theta = v^2/r$, where θ is the angle of bank required for the turn.

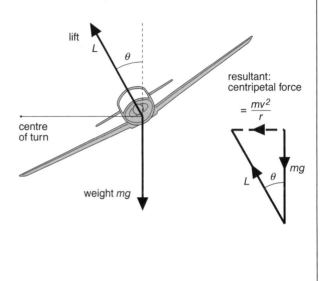

4.2.2 Gravitational fields

Gravitational force

All masses attract each other with a **gravitational force**. If point masses M and m, are a distance r apart, and F is the force on each, then according to **Newton's law of gravitation**

$$F \propto \frac{Mm}{r^2}$$

With a suitable constant, the above proportion can be turned into an equation:

$$F = -\frac{GMm}{r^2}$$

G is called the **gravitational constant**. It is found by experiment using large laboratory masses and an extremely sensitive force-measuring system. In SI units, the value of G is 6.67×10^{-11} N m² kg⁻².

The minus sign is because the force between the masses is an attractive force.

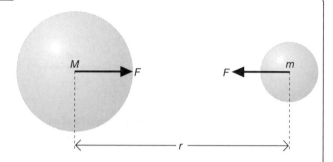

The equation on the left is also valid for spherical masses of uniform density, with centres r apart, as above.

Note:
- Newton's law of gravitation is an example of an **inverse square law**. If the distance r doubles, the force F drops to one quarter, and so on.
- Gravitational forces are always forces of *attraction*.
- Gravitational forces are extremely weak, unless at least one of the objects is of planetary mass or more.

Example

The gravitational force between the Sun and the Earth. (Sun's mass = 1.99×10^{30} kg, Earth's mass = 5.98×10^{24} kg, Earth to Sun distance = 1.50×10^{11} m)

$$F =$$

$$\frac{-(6.67 \times 10^{-11} \text{ N m}^2 \text{ kg}^{-2}) \times (1.99 \times 10^{30} \text{ kg}) \times (5.98 \times 10^{24} \text{ kg})}{(1.50 \times 10^{11} \text{ m})^2}$$

$$= 3.53 \times 10^{22} \text{ N}$$

Gravitational field

If a mass feels a gravitational force, then it is in a **gravitational field**.

> The **gravitational field strength** g is the force per unit mass.

$$g = \frac{\text{gravitational force}}{\text{mass}} \qquad \text{In symbols } g = \frac{F}{m}$$

For example, if a mass of 2.0 kg feels a gravitational force of 10 N, then g is 5.0 N kg⁻¹.

Note:
- Gravitational field strength is a vector.
- g is a variable and can have different values. The symbol g above does not imply the particular value of 9.81 N kg⁻¹ near the Earth's surface.
- The force acting on a mass in a gravitational field can be found by rearranging the equation above: $F = mg$

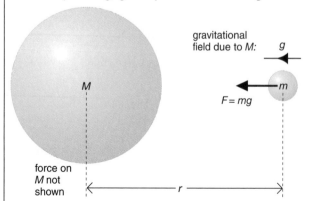

gravitational field due to M: g

$F = mg$

force on M not shown

Above, mass M produces a gravitational field which acts on mass m.

As $\quad F = -\frac{GMm}{r^2} \quad$ and $\quad F = mg$

it follows that $\qquad g = -\frac{GM}{r^2}$

It is equally true to say that m produces a gravitational field which acts on M. Either way, mass × gravitational field strength gives a force of the same magnitude, $-GMm/r^2$.

Example

Estimate the gravitational field strength on Jupiter, assuming the mass is concentrated at a point at the centre of the planet. Jupiter's mass = 1.90×10^{27} kg, radius = 7.14×10^7 m

$$g = -\frac{(6.67 \times 10^{-11} \text{ N m}^2 \text{ kg}^{-2}) \times (1.90 \times 10^{27} \text{ kg})}{(7.14 \times 10^7 \text{ m})^2}$$

$$= 24.9 \text{ N kg}^{-1}$$

Gravitational field *continued*

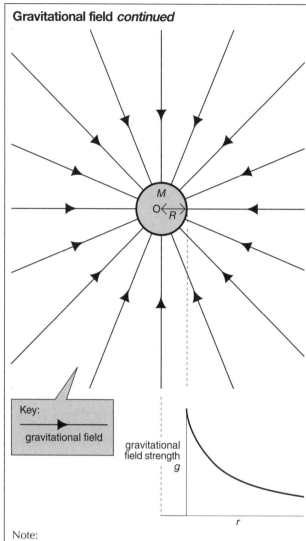

Key:

———▶

gravitational field

gravitational field strength g

r

Note:
- The gravitational field around a spherical mass is shown above. It is called a **radial** field because of its shape. Inside the mass, the equation on the left does not apply. g falls to zero at the centre.

The Earth's gravitational field strength

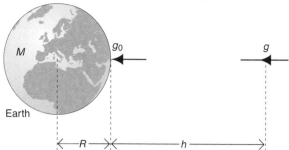

Earth

$\leftarrow R \rightarrow$ $\leftarrow\quad h\quad\rightarrow$

At the surface If M is the Earth's mass, R is its radius, and g_0 is the gravitational field strength at its surface, then

$$g_0 = -\frac{GM}{R^2} \qquad (1)$$

Note:
- g_0 is 9.81 N kg^{-1}. It is more commonly known as g (without the $_0$). Here however, the $_0$ has been added to distinguish it from other possible values of g.
- Using measured values of g_0, R, and G in the above equation, the Earth's mass M can be calculated. With R known, the Earth's average density can also be found.

Above the surface In this case, $g = -GM/r^2$. From this and equation (1), the following result is obtained:

$$g = \frac{g_0 R^2}{r^2}$$

So as the distance from the Earth increases, g decreases.

Close to the Earth's surface
The mean radius of the Earth is 6.378×10^6m, so 1000m above the Earth's surface:

$$g = \frac{g_0 R^2}{r^2} = (9.81 \text{ N kg}^{-1}) \times \frac{(6.378 \times 10^6 \text{ m})^2}{(6.378 \times 10^6 \text{ m} + 1000 \text{ m})^2}$$

$$g = (9.81 \text{ N kg}^{-1}) \times 0.9996 = 9.81 \text{ N kg}^{-1}$$

So close to the Earth's surface the field is uniform (to at least 3 sf) and approximately equal to the acceleration of free fall (9.81 m s^{-2})

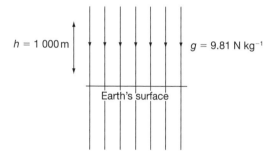

$h = 1\,000\,\text{m}$ $g = 9.81$ N kg^{-1}

Earth's surface

Note:
- the gravitational force attracting a 1kg mass towards the centre of the Earth is the same as the acceleration on the 1 kg mass.
- Even at a height of 10 000 m the calculated value is 9.78 N kg^{-1} The Earth is not a perfect sphere and this also leads to a slight variation in g.

An orbit equation

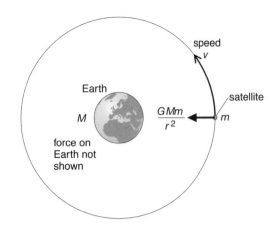

speed
v

Earth

M

$\dfrac{GMm}{r^2}$

satellite
m

force on Earth not shown

Above, a satellite is in a circular orbit around the Earth. The gravitational force on the satellite provides the centripetal force needed for the circular motion. So

$$\frac{GMm}{r^2} = \frac{mv^2}{r} \qquad (2)$$

Note:

- The equation can be used to find the speed v needed for an orbit of any given radius r.
- As m cancels from each side of the equation, the speed needed does not depend on the mass of the satellite.

A value for GM To use the above equation, you do not need to know the Earth's mass M. Instead, a value for GM can be found using equation (1). Rearranged, this gives

$$GM = g_0 R^2 \qquad (3)$$

where R is the Earth's radius (6.37×10^6 m), and g_0 is the gravitational field strength at its surface (9.81 N kg^{-1}).

Period of orbit

The period T is the time taken for one orbit.

So, as shown in 4.2.1. Circular motion (page 22), speed $v = \dfrac{2\pi r}{T}$

Substituting in equation (2) left

$$\frac{GMm}{r^2} = m \,\frac{(2\pi r)^2}{T^2}$$

rearranging gives the following link between T and r.

$$T^2 = \left(\frac{4\pi^2}{GM}\right) r^3 \qquad (4)$$

As $4\pi^2/GM$ is a constant, T^2/r^3 has the same value for all satellites. So as r increases, the period gets longer.

Example

The planet Mercury orbits the Sun at a mean distance of 5.79×10^{10} m. Assuming its orbit is circular, what is the period? (The mass of the Sun is 1.99×10^{30} kg)

Using equation (4)

$$T^2 = \frac{4\pi^2 \times (5.79 \times 10^{10}\text{ m})^3}{(6.67 \times 10^{-11}\text{N m}^2\text{ kg}^{-2}) \times (1.99 \times 10^{30}\text{ kg})}$$

$T = \sqrt{(5.77 \times 10^{13}\text{ s}^2)} = 7.6 \times 10^6$ s = 88 days

'Weightlessness'

An astronaut in a satellite is in a state of free fall. Her acceleration towards the Earth is exactly the same as that of the satellite, so the floor of the satellite exerts no forces on her. As a result, she experiences exactly the same sensation of weightlessness as she would in zero gravity. However, she is not really weightless. A few hundred kilometres above the Earth, the gravitational force on her is almost as strong as it is down on the surface.

Kepler's third law

Johannes Kepler analysed data on the orbits and positions of the planets and found that:

$$T^2 \propto r^3 \qquad (5)$$

Where T is the period of the orbit around the Sun, and r is the mean distance from the centre of the Sun. Kepler did not know why this was true – his law was an empirical Law (one based on observations.) Later, when Isaac Newton published his work, one of the reasons for accepting his Theory of Gravitation was that it explained Kepler's third law.

Example

The orbital period of the Earth is 365.3 days and the mean radius of its orbit is 1.496×10^{11} m. Mars is 2.28×10^{11} m from the Sun, what is its orbital period?

Using equation (5)

T^2 is a constant, so for Mars:

$$T^2 = \frac{(365.3 \text{ days})^2}{r^3} \times \frac{(2.279 \times 10^{11}\text{ m})^3}{(1.496 \times 10^{11}\text{ m})^3}$$

$T = \sqrt{(4.718 \times 10^5)}$ days = 687 days

	Distance from Sun/ $\times 10^7$ km	Period of orbit/ days
Mars	22.8	687.0
Earth	15.0	365.3
Venus	10.8	224.7
Mercury	5.8	88.0
Sun		

Geostationary orbit

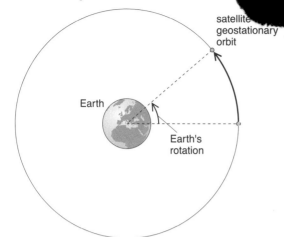

Earth

Earth's rotation

satellite geostationary orbit

If a satellite is in a **geostationary orbit**, then viewed from Earth, it appears to be in a fixed position in the sky. This is because the period of its orbit exactly matches the period of the Earth's rotation (24 hours). Communications satellites are normally in geostationary orbits.

$r^3 = 7.54$

$r = 4.22 \times 10^7$ m

The radius of the Earth is $6.39 \times$ the surface of the Earth is

4.22×10^7 m $- 6.39 \times 10^6$ m $= 3.6 \times 10^7$ m

Note that the satellite will need to be placed at above the equator if it is to stay in the same position the Earth's surface.

Uses of geostationary satellites
The advantage is that ground stations communicating with the satellite do not have to track it across the sky.

- Telecommunications, especially TV.
- Some meteorological satellites and other Earth monitoring satellites.

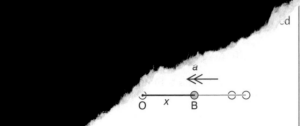

above shows a particle B oscillating about O
~~~M. The position of B is shown at equal intervals of
~~~. Its acceleration is proportional to its displacement from
O, and always directed towards O.

If x is the displacement, and a is the acceleration (in the x direction), then this can be expressed mathematically:

$$a = -(\text{positive constant})\, x \qquad (1)$$

The minus sign indicates that a is always in the opposite direction to x.

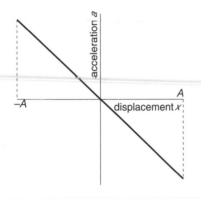

Periodic motion
This is motion in continually repeating **cycles**.

T is the **period** (the time for one cycle).
f is the **frequency** (the number of cycles per second).
ω is the **angular frequency**, the angle swept out by a radial line per second (measured in rad s^{-1}).

T, f, and ω are linked by the equations below:

$$T = \frac{1}{f} \qquad\qquad \omega = 2\pi f \qquad\qquad T = \frac{2\pi}{\omega}$$ (see page 23)

Displacement, x is the distance of an object from its mean, or rest position. It is a vector quantity (can be positive or negative) and is measured in metres.

Amplitude, A is the maximum displacement, measured in metres.

Examples of free oscillations
A swinging pendulum will eventually stop due to friction and air resistance so it is not a completely free oscillation. If there was no friction or air resistance it would keep swinging for ever. The movement of the mass would be a free oscillation. The graph shows how its position changes with time.

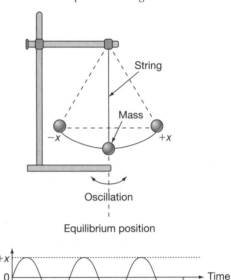

If a mass on a spring is pulled down slightly and released it will oscillate about its equilibrium position. Again, if there was no friction or air resistance this would be a free oscillation. The same graph shows how its position changes with time.

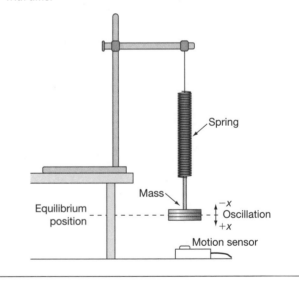

Equations for SHM

Acceleration:

When a particle is moving in SHM with frequency f, the equation for SHM can be written:

$$a = -(2\pi f)^2 x$$

Displacement:

When A is the amplitude of the oscillation, two possible solutions to this equation, which show how the displacement, x, varies with time, t, are:

$x = A \sin(2\pi ft)$ for a particle where $t = 0$ occurs at the rest position, $x = 0$. ($\sin 0 = 0$)

and

$x = A \cos(2\pi ft)$ for a particle where $t = 0$ occurs at the position of maximum amplitude, $x = A$. ($\cos 0 = 1$)

Example

A particle moves with SHM. The period is 2.0 s and the amplitude is 5.0 cm. What is its displacement 0.25 s after passing through rest position?

Using $x = A \sin(2\pi ft)$ because $x = 0$ when $t = 0$

$$f = \frac{1}{T} = \frac{1}{(2.0s)} = 0.5 \text{ Hz}$$

$x = (5.0 \text{ cm}) \sin (2\pi \times 0.5 \text{ Hz} \times 0.25 \text{ s}) = (5.0 \text{ cm}) \sin (0.25\pi)$

remember that the angle is in radians so that you must either convert it to degrees or switch your calculator to radians.

$x = (5.0 \text{ cm}) \times 0.71 = 3.6 \text{ cm}$

Velocity:

The velocity will be zero when the displacement is a maximum ($x = \pm A$) because at this point the particle stops and changes direction, moving back towards the rest position.

The velocity will be a maximum value as the particle passes through the rest position.

The graph of acceleration against displacement in 'Defining simple harmonic motion', page 30, shows that the acceleration towards the rest position drops to zero as the particle reaches the rest position and then becomes negative, slowing the particle down, so the maximum speed occurs at the rest position.

$$v_{max} = (2\pi f)A$$

Example

What is the maximum speed of the particle in the above example?

$$v_{max} = (2\pi \times 0.5 \text{Hz}) \times 5 \text{ cm} = 16 \text{ cm s}^{-1}$$

Note:

- These equations can also be written in terms of the angular frequency, ω where $\omega = 2\pi f$.

  $a = -\omega^2 x$

  $x = A \sin \omega t$ and $x = A \cos \omega t$

  $V_{max} = \omega A$

Displacement–time graph for SHM

The following graph shows how the displacement varies with time for one complete oscillation starting from the centre of oscillation.

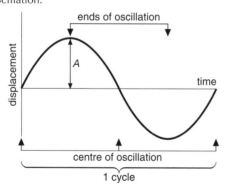

A is the **amplitude** of the oscillation.

Phase difference

An oscillation that has the same period but reaches its peak at a different time to that shown above is said to have a **phase difference**. It is quoted as an angle in radians not a time. Two important examples are shown below:

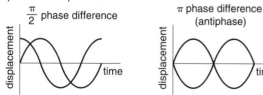

Linking circular motion and SHM

Below, particle P is moving in a circle with a steady angular velocity ω. Particle B is oscillating about O along the horizontal axis so that it is always vertically above or beneath P. The amplitude A of the oscillation is equal to the radius of the circle, r.

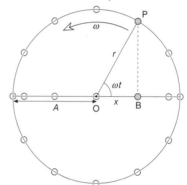

For particle B, $x = r \cos \omega t$ (2)

Using calculus, B's velocity v and acceleration a can be found from the above equation. These are the results:

$$v = -r\omega \sin \omega t \quad (3)$$

$$a = -r\omega^2 \cos \omega t \quad (4)$$

From equations (4) and (2), it follows that

$$a = -\omega^2 x \quad (5)$$

From the diagram, $\sin \omega t = \dfrac{\sqrt{r^2 - x^2}}{r}$

Using equation (3) and remembering that $r = A$, the **amplitude** of the oscillation, the velocity at a distance x from the centre of oscillation can be calculated from

$$v = \omega\sqrt{A^2 - x^2} = 2\pi f\sqrt{(A^2 - x^2)} \quad (6)$$

$$v_{max} = \omega A \text{ (when } x = 0)$$

$$a_{max} = -\omega^2 A \text{ (when } x = A)$$

Note:

- Equation (5) has the same form as equation (1). So particle B is moving with SHM.
- The constant in equation (1) is equal to ω^2.
- Using calculus notation, the equation for SHM can be written in the following form:

$$\frac{d^2x}{dt^2} = -\omega^2 x$$

SHM and a mass on a spring

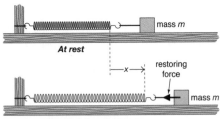

At rest

After stretch and release

Above, a mass is attached to a spring. When pulled and released, the mass makes small oscillations about the equilibrium point.

The *spring constant* or *stiffness of the spring*, k is given by:

$$k = \frac{F}{x}$$

So, if the mass m is pulled by x and then released,

$$\text{resultant force on mass} = kx$$

But, $\qquad\qquad$ force = mass × acceleration

So, $\qquad\qquad$ $\text{acceleration} = \dfrac{kx}{m}$

So, $\quad$ acceleration (in x direction) $a = -\dfrac{kx}{m}$

Comparing this with equation (5) shows that the motion is SHM and that

$$\frac{k}{m} = \omega^2$$

As $\quad T = \dfrac{2\pi}{\omega}$ $\qquad$ $\boxed{T = 2\pi\sqrt{\dfrac{m}{k}}}$

In any oscillating system to which Hooke's law applies, the motion is SHM.

SHM and the simple pendulum

Provided its swings are small, and air resistance is neglible, a simple pendulum moves with SHM. The following analysis shows why.

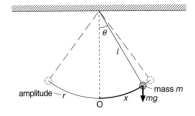

The mass m (above) has been displaced by x. It is being pulled towards O by a component of its weight:

$$\text{force (towards O)} = mg \sin \theta$$

But for very small angles $\qquad \sin \theta = \dfrac{x}{l}$

So $\qquad\qquad$ force (towards O) $= \dfrac{mgx}{l}$

But $\qquad$ force = mass × acceleration

So $\qquad\qquad$ acceleration (towards O) $= \dfrac{gx}{l}$

So $\qquad$ acceleration (in x direction) $a = \dfrac{-gx}{l}$

Graphs of SHM

The graphs below are for an object moving with SHM: for example, a pendulum making small swings.

- At time = 0 the pendulum bob starts at its maximum displacement from point O, the equilibrium position. Its initial velocity is zero, and it has a maximum acceleration towards O.
- The velocity increases and the acceleration and displacement decrease as the pendulum bob moves towards O.
- When it reaches O, the velocity is a maximum, and the acceleration and displacement are both zero.
- The pendulum bob passes through O and the acceleration and displacement start increasing. The acceleration is still towards O so the velocity is now decreasing.
- When the pendulum bob reaches its maximum displacement on the other side of the equilibrium position, half way through the cycle, its velocity has slowed to zero and the acceleration is a maximum towards O.
- The velocity begins to increase in the direction of O and the acceleration and displacement decrease.
- When it reaches O again, the velocity is a maximum again, but in the opposite direction, and the acceleration and displacement are again both zero.
- The acceleration increases towards O and the velocity decreases until the pendulum bob reaches the maximum displacement, back at the starting position. The velocity is once again zero, and the acceleration is a maximum towards O.

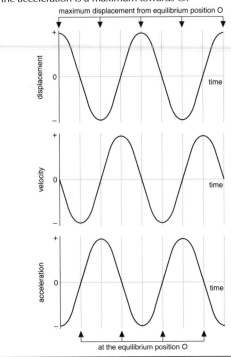

Comparing this with equation (5) shows that the motion is SHM and that

$$\frac{g}{l} = \omega^2$$

As $\quad T = \dfrac{2\pi}{\omega}$ $\qquad$ $\boxed{T = 2\pi\sqrt{\dfrac{l}{g}}}$

Note:
- Another characteristic of SHM is that the period of motion is independent of the amplitude, so if a pendulum swings higher or a mass on a spring is pulled further before release the period of the oscillations is unchanged.

Energy in SHM

For a system with free oscillations, no energy is transferred to the surroundings, the amplitude stays the same and the graph of the motion against time is a sine wave pattern.

During the motion the energy is transferred between kinetic and potential energy. The maximum kinetic energy (E_k) is at the equilibrium or rest position, and the minimum is when the displacement $x = \pm A$ where A is the amplitude. The maximum potential energy (E_p) occurs where $x = \pm A$ and the minimum is at the rest position.

For a mass on a spring system the E_p is elastic E_p, and for a pendulum the E_p is gravitational E_p.

Mass–spring system

The elastic (or strain energy) stored in a mass–spring system is the work done in stretching the spring. This is the area under the force–displacement graph.

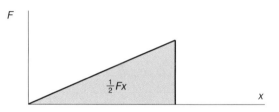

$$\text{elastic energy stored} = \tfrac{1}{2} Fx$$

Since $F = kx$, another useful equation for stored energy is

$$\text{elastic energy stored} = \tfrac{1}{2} kx^2$$

As the spring moves toward the equilibrium position it loses elastic stored energy and gains kinetic energy. But in the absence of any damping the total energy remains constant.

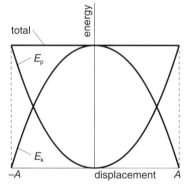

$$\text{maximum elastic stored energy} = \tfrac{1}{2} kA^2$$
$$\text{maximum kinetic energy} = \tfrac{1}{2} m(2\pi f A)^2$$

Damping

A mass–spring system or a pendulum will not go on swinging for ever. Energy is gradually lost to the surroundings due to air resistance or some other resistive force and the oscillations die away. This effect is called **damping**.

In road vehicles, dampers (wrongly called 'shock absorbers') are fitted to the suspension springs so that unwanted oscillations die away quickly. Some systems (for example moving-coil ammeters and voltmeters) have so much damping that no real oscillations occur. The minimum damping needed for this is called **critical damping**.

Pendulum

For a pendulum the mass loses potential energy (E_p) as it swings downwards and gains kinetic energy (E_k).

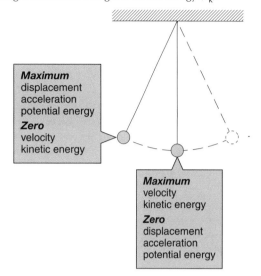

Maximum
displacement
acceleration
potential energy
Zero
velocity
kinetic energy

Maximum
velocity
kinetic energy
Zero
displacement
acceleration
potential energy

If there is no air resistance the total $E_p + E_k$ is constant.
$$\text{total energy} = \text{maximum } E_k = \tfrac{1}{2} m(2\pi f A)^2$$

Notice that in all the equations for total energy the total energy is proportional to the *square* of the amplitude (A^2).

This graph shows how the energy changes with time for a pendulum swinging.

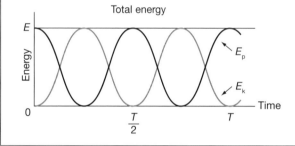

The rate at which the amplitude falls depends on the fraction of the existing energy that is lost during each oscillation.

In a **lightly damped** system only a small fraction is lost so that the amplitude of one oscillation is only slightly lower than the one before.

The graphs here are for oscillations with different degrees of damping.

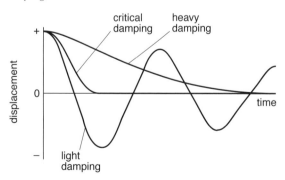

Other uses for dampers include fitting them to buildings in earthquake zones to reduce vibrations so the building can withstand earthquakes. The Millennium Bridge in London has dampers fitted to stop it oscillating from side to side.

Natural frequency

This is the frequency of the oscillation that occurs when the mass of an oscillator (such as a pendulum bob or mass of a mass–spring system) is displaced and then released. The only forces acting are the internal forces of the oscillating system. These oscillations are **free oscillations**.

Forced oscillations

Forced oscillations occur when an external periodic force acts on an object that is free to oscillate.

Examples include:
- engine vibrations making bus windows oscillate
- the spinning drum causing vibrations in a washing machine
- the body of a guitar vibrating when a string is plucked.

The body that is forced to oscillate vibrates at the same frequency as that of the external source that is providing the energy.

The amplitude of the oscillations produced depends on
- how close the external frequency is to the natural frequency of the oscillator
- the degree of damping of the oscillating system.

Resonance

Resonance occurs when the frequency of the external source that is driving the oscillation is equal to the natural frequency of the oscillator that is being driven into oscillations.

When resonance occurs the amplitude of the resulting oscillations is a maximum.

The following graph shows how the amplitude of an oscillator with natural frequency f_0 varies with the frequency of the frequency of the source that is driving the oscillations.

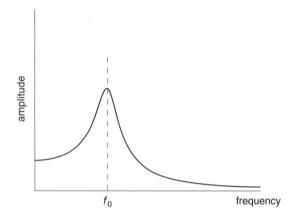

Graphs such as these are called *frequency response graphs*.

Effect of damping on resonance

When a system is lightly damped, it loses very little energy during an oscillation. If it is being forced to oscillate it will retain most of the energy put into it so that the energy stored builds up and the amplitude becomes very large.

When a system is heavily damped, energy is lost quickly so that the amplitude is lower.

The graph shows the frequency response for lightly and heavily damped oscillators that have the same natural frequency.

Note:
- The amplitude of an oscillation stops increasing when the energy put in each cycle is equal to that lost during the cycle.

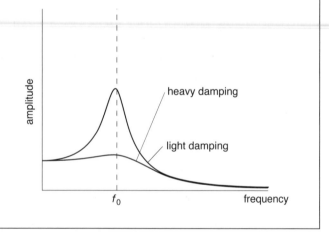

Examples of resonance

Useful resonance

In microwave ovens, the microwaves are produced by a cavity magnetron. There is a resonant frequency that depends on the size of the cavity.

The microwaves are emitted at the resonant frequency and directed into the oven. The frequency is chosen to be one that vibrates the water, fat, and sugar molecules, which heats the food.

Resonance is also useful in musical instruments, for example the air in a clarinet when the reed vibrates.

Electrical resonance

Some electrical circuits (you do not need to know the details) can also be forced into oscillations or be made to resonate.

This effect is used in the tuner of a radio receiver. The current in the circuit reaches a maximum at the resonant frequency. By tuning the circuit so that the natural frequency matches the frequency of the radio channel, only the frequencies in a small range around this frequency are selected.

In MRI (magnetic resonance imaging) the nuclei of atoms are made to oscillate in a magnetic field. The radio frequency radiation emitted is used to form an image.

Resonance to be avoided

Suspension bridges must be designed so that resonance is not caused by the wind vibrating the bridge. The Tacoma Narrows Bridge in Washington State, USA collapsed because of this effect.

Module 3: Thermal physics

4.3.1 Solid, liquid and gas

Solids, liquids, and gases

According to the *kinetic theory*, matter is made up of tiny, randomly moving particles. Each particle may be a single atom, a group of atoms called a *molecule*, or an ion (but for convenience, from now on we will refer to molecules). The three normal *phases* of matter are solid, liquid, and gas.

Solid The molecules are held close together by strong forces of attraction. They vibrate, but about fixed central positions, so a solid keeps a fixed shape and volume.

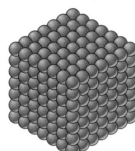
Solid

The centres of the molecules are typically about a few tenths of a nanometre (3×10^{-10} m) apart. The shapes of crystals result from the way the ions join together, for example table salt (sodium chloride) has a cubic structure, and pure crystals are cubes. Assuming the molecules are spheres, when they are packed as closely as possible then a pyramid shape is formed. This is called close-packing.

Close packing

Liquid The molecules are held close together. But the vibrations are strong enough to overcome the attractions, so the molecules can change positions. A liquid has a fixed volume, but it can flow to fill any shape.

The molecules are almost as close together as solids – usually a solid expands slightly as its temperature rises and so does a liquid. In the case of water the molecules are closer together as a liquid at 0° C than as a solid, because of the way the molecules are arranged in the solid. The densities of solids and liquids are similar and the spacing is a few tenths of a nanometre.

Liquid

Gas The molecules move at high speed, colliding with each other and with the walls of their container. They are too spread out and fast-moving to stick together, so a gas quickly fills any space available. Its pressure is due to the impact of its molecules on the container walls.

The densities of gases at atmospheric pressure are about a thousand times less than solids and liquids, so the molecules are about ten times further apart. (In a solid cube there are $10 \times 10 \times 10$ molecules = 1000 but in the same size cube of gas there would be, on average, only one molecule of gas.)

Note:
- Diagrams showing the difference should show that the molecules do not change size and in liquids they are not much more widely spaced – but in gases they are.

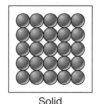
Solid

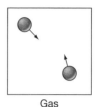
Liquid

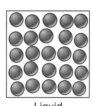
Gas

Brownian motion

When very small particles such as smoke in air, or pollen in water, are viewed through a microscope, they appear to be tiny specks of light shimmering on a grey background. Closer inspection with a higher powered microscope shows that the particle is moving in a zig-zag path and changing direction in a completely random pattern.

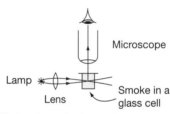

The cell is filled with smoke from a burning waxed paper drinking straw. It is illuminated and viewed through the microscope. The motion is called Brownian motion after Robert Brown who saw the effect with pollen grains. The changes in direction are caused by fluctuations in the thousands of simultaneous collisions between the air molecules and the smoke particle. This causes a resultant force on the particle for an instant. The air molecules are too small to see but the resultant force is large enough to change the direction of the larger, and more massive, smoke particle.

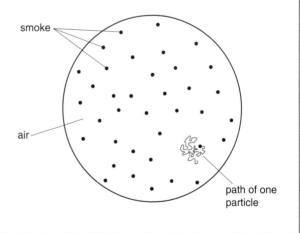

Pressure

> Pressure is the force per unit area, where the force is perpendicular to the area.

$$\text{pressure } (p) = \frac{\text{perpendicular force } (F)}{\text{area } (A)}$$

Pressure is measured in pascals (Pa).

$1 \text{ Pa} = 1 \text{ Nm}^{-2}$

Pressure in gases

The kinetic model can be used to describe gas pressure. In the model gases are made up of molecules which are moving quickly and colliding with the walls of the container and with each other. It is the collisions with the walls of the container that give the pressure (not collisions between molecules.) The more gas there is in the container, the more collisions and the higher the pressure. The higher the temperature of the gas the faster the molecules will be moving so the greater their kinetic energy and the greater the force on the walls of the container.

The kinetic model and change of state

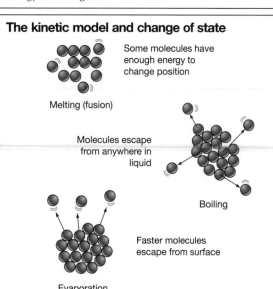

Some molecules have enough energy to change position

Melting (fusion)

Molecules escape from anywhere in liquid

Boiling

Faster molecules escape from surface

Evaporation

Melting When the molecules in a solid gain enough potential energy to break the bonds holding them in fixed positions relative to each other, they are then able to move over and around each other and have become a liquid.

Boiling When the molecules anywhere in the liquid gain enough potential energy to break the bonds holding them to other molecules completely and leave the liquid. The have become a gas.

Evaporation When a liquid evaporates, molecules escape from its surface and move about freely as a gas.

In a liquid, the vibrating molecules keep colliding with each other, some gaining kinetic energy and others losing it. At the surface, some of the faster, upward-moving molecules have enough kinetic energy to overcome the attractions from other molecules and escape from the liquid. With these faster molecules gone, the average E_k of those left behind is reduced i.e. the temperature of the liquid falls. That is why evaporation has a cooling effect.

The rate of evaporation (and therefore the rate at which heat is lost from a liquid) is increased if:
- the surface area is increased (more of the faster molecules are near the surface),
- the temperature is increased (more of the molecules have enough kinetic energy to escape),
- the pressure is reduced (escaping molecules are less likely to rebound from other molecules back into the liquid),
- there is a draught across the surface (escaping molecules are removed before they can rebound),
- gas is bubbled through the liquid.

Internal energy

The molecules in a solid have kinetic energy due to their vibrations. In liquids and gases they also have kinetic energy due to translational motion. The molecules also have potential energy.

> The **internal energy** of a system is the sum of the random distribution of kinetic and potential energies associated with the molecules of the system.

Internal energy and change of state

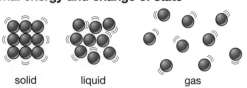

solid   liquid   gas

In the solid, the molecules are vibrating about equilibrium positions. A molecule has maximum kinetic energy (E_k), and minimum potential energy (E_p), when it is in its equilibrium position. When the molecule is at its maximum displacement from equilibrium, its E_k is zero and its E_p is a maximum. The total of all the E_k and E_p for all the molecules is equal to the **internal energy** of the solid.

When the solid changes to a liquid the temperature does not change, so the average E_k of the molecules is unchanged. The molecules have enough energy to move apart, against the forces of attraction between them, and change their positions within the liquid. This is an increase in E_p, so the internal energy of the solid increases as it melts to a liquid.

When the liquid becomes a gas, the temperature is the same as before. So the average E_k of each particle due to its linear motion is the same. However, the average E_p is more because of the increased separation of the molecules. The gas has more internal energy than the liquid.

Internal energy and temperature

The molecules in, for example, a gas move at a range of speeds. However, the higher the temperature, the faster the molecules move on average.

Molecular speeds in a gas

In any gas, the molecules randomly collide with each other. In these collisions, some molecules gain energy (and therefore speed) while others lose it. As a result, at any instant, the molecules have a range of speeds, as shown in the distribution graph below.

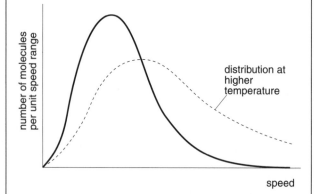

If two objects at the same temperature are in contact, there is no flow of heat between them. This is because the average kinetic energy of each particle due to its vibrating or speeding motion is the same in each object, so there is no overall transfer of energy from one object to the other.

4.3.2 Temperature

Temperature and thermal equilibrium

Objects A and B are in contact. If net thermal energy flows from A to B, then A is at a higher **temperature** than B.

When the net thermal energy flow from A to B is zero, the two objects are in **thermal equilibrium** and at the same temperature.

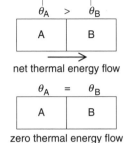

Thermodynamic temperatures

The **Kelvin scale** is a thermodynamic scale, related to the average kinetic energy per molecule.

On the Kelvin scale **absolute zero** is 0 kelvin (0 K). This is the temperature at which all substances have minimum internal energy. One advantage of using this scale is that the temperature in kelvin is proportional to the average kinetic energy per molecule.

Temperature scales

Celsius scale On this scale, pure water freezes at 0 °C and boils at 100 °C (under standard atmospheric conditions).

Kelvin scale One kelvin (K) is the same size as one degree in the Celsius scale, but its 'zero' is **absolute zero** (−273.15 °C), the temperature at which particles have the minimum internal energy.

$$T(K) = 0 \ (°C) + 273.15 \, K$$

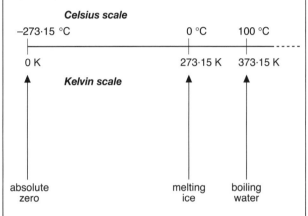

Note:
- Using 273 instead of the value of 273·15 gives enough accuracy in most cases.

4.3.3 Thermal properties of materials

Heat capacity

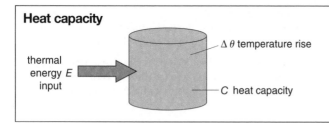

thermal energy E input

$\Delta \theta$ temperature rise

C heat capacity

The **heat capacity** of an object is given by this equation:

$$\text{heat capacity} = \frac{\text{heat input}}{\text{temperature rise}} \qquad C = \frac{E}{\Delta\theta}$$

For example, if a heat input of 4000 J causes a temperature rise of 2 K, then the heat capacity is 2000 J K^{-1}.

Specific heat capacity

The heat capacity per unit mass is called the **specific heat capacity**. If a substance's specific heat capacity is c, then, for a mass m:

$$E = mc\Delta\theta \qquad (1)$$

Water has a high specific heat capacity (4200 J kg^{-1} K^{-1}). This makes it a good 'heat storer'. A relatively large thermal energy input is needed for any given temperature rise, and there is a relatively large thermal energy output when the temperature falls.

Measuring c for a liquid (e.g. water) This can be done using the equipment below. The principle is to supply a measured mass of liquid with a known amount of thermal energy from an electric heating coil, measure the temperature rise, and calculate c using equation (1).

low voltage DC supply

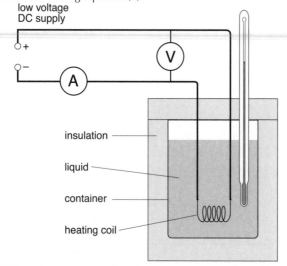

insulation

liquid

container

heating coil

If the p.d. across the coil is V, and a current I passes for time t, then the electrical energy supplied = VIt.
If all this energy is supplied as thermal energy, and none is lost:

$$VIt = mc\Delta\theta$$

Knowing the mass m and temperature rise $\Delta\theta$ of the liquid, its specific heat capacity c can be calculated.

Note:
- When the water heats up, its container does as well. For greater accuracy, this must be allowed for.
- Some thermal energy is lost, despite the insulation. However, there are experiments in which, using different sets of results, heat losses can be eliminated from the calculation.

Measuring c for a solid (e.g. a metal) The method is essentially the same as that shown above, except that a solid block is used instead of the liquid. The block has holes drilled in it for an electric heater and a thermometer.

Copper has a **specific heat capacity** of 390 J kg^{-1} K^{-1}. This means that 390 J of energy are required to raise the temperature of 1 kg of copper by 1 K. To raise the temperature of 2 kg of copper by 10 K, the heat input required = $2.0 \times 390 \times 10 = 7800$ J.

Changing state

The graph shows what happens when a very cold solid (ice) takes in thermal energy at a steady rate. Melting and boiling are both examples of a change of **state**.

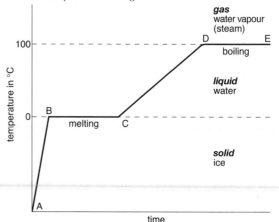

A to B The temperature rises until the ice starts to melt.

B to C Thermal energy is absorbed, but with no rise in temperature. The energy input is being used to overcome the attractions between the molecules as the solid changes into a liquid.

C to D The temperature rises until the water starts to boil.

D to E Thermal energy is absorbed, but with no rise in temperature. The energy input is being used to separate the molecules as the liquid changes into a gas (water vapour).

Boiling is a rapid type of evaporation in which vapour bubbles, forming in the liquid, expand rapidly because their pressure is high enough to overcome atmospheric pressure.
The thermal energy required to change a liquid into a gas (or a solid into a liquid) is called **latent heat**.

Specific latent heat of vaporization

The **specific latent heat of vaporization** of a substance is the thermal energy which must be supplied per unit mass to change a liquid into a gas, without change in temperature.
If E is the thermal energy supplied, m is the mass, and l_v is the specific latent heat of vaporization, then

$$E = ml_v$$

The specific latent heat of vaporization of water is 2.3×10^6 J kg^{-1}. So, to turn 2.0 kg of water into water vapour (at the same temperature) would require 4.6×10^6 J of thermal energy.

Specific latent heat of fusion

The **specific latent heat of fusion** of a substance is the thermal energy which must be supplied per unit mass to change a solid into a liquid, without change in temperature.
If E is the thermal energy supplied, m is the mass, and l_f is the specific latent heat of fusion, then

$$E = ml_f$$

The specific latent heat of fusion of water is 3.3×10^5 J kg^{-1} (about $\frac{1}{7}$ of its specific latent heat of vaporization).

4.3.4 Ideal gases

Boyle's law

Experiments show that, for a fixed mass of gas at constant temperature, the pressure p decreases when the volume V is increased. A graph of p against V for air is shown on the right.

According to **Boyle's law**, for a fixed mass of gas at constant temperature,

$$pV = \text{constant}$$

If $pV = \text{constant}$, then $p \propto 1/V$. So a graph of p against $1/V$ is a straight line through the origin.

Note:
- The value of the constant depends on the mass of the gas and on its temperature. The dashed lines show the effects of raising the temperature.
- Under some conditions, the behaviour of real gases departs from that predicted by Boyle's law (see below).

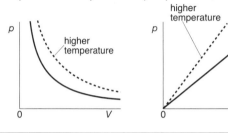

Charles's law

For a fixed mass of gas at constant pressure, the volume V increases with the Kelvin temperature T, as on the right.

According to **Charles's law**, for a fixed mass of gas at constant pressure,

$$\frac{V}{T} = \text{constant}$$

From this equation, $V \propto T$, which is why the graph is a straight line through the origin.

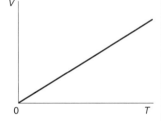

Note:
- Charles's law predicts zero volume at absolute zero. However no real gas behaves like an ideal gas at near-zero volume and temperature.

The pressure law

Before reading this panel, see I1 on temperature.

According to the **pressure law**, for a fixed mass of gas at constant volume,

$$\frac{p}{T} = \text{constant}$$

From this equation, $p \propto T$, so the graph of p against I is a straight line through the origin, as shown.

Note:
- The pressure law predicts that the pressure of any ideal gas should be zero at absolute zero. This concept is used to define the zero point (0 K) on the Kelvin scale and to find its Celsius equivalent ($-273.15\ ^\circ$C).

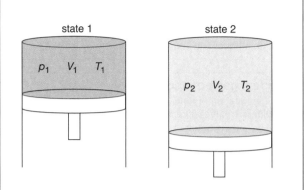

Equation of state

If a fixed mass of gas changes from state 1 to state 2, at a different pressure, volume, and kelvin temperature, then

$$\frac{pV}{T} = \text{constant}$$

or $\dfrac{p_1 V_1}{T_1} = \dfrac{p_2 V_2}{T_2}$

This is called the **equation of state** for an ideal gas.

Note, in the above equation:
- If $T_1 = T_2$, then $p_1 V_1 = p_2 V_2$. This is Boyle's law.

- If $V_1 = V_2$, then $\dfrac{p_1}{T_1} = \dfrac{p_2}{T_2}$.

- If $p_1 = p_2$, then $\dfrac{V_1}{T_1} = \dfrac{V_2}{T_2}$.

Kinetic theory for an ideal gas

The laws governing the behaviour of ideal gases can be deduced mathematically from the kinetic theory. In using the theory, the following assumptions are made:
- The motion of the molecules is completely random.
- The forces of attraction between the molecules are negligible.
- The molecules themselves have a negligible volume compared with the volume occupied by the gas.
- The molecules make perfectly elastic collisons with each other and with the walls of their container.
- The number of molecules is so large that there are billions of collisions per second.
- Each collision takes a negligible time.
- Between collisions, each molecule has a steady speed.

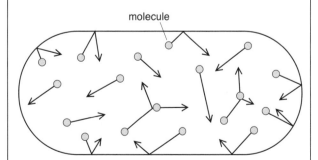

The diagram above shows a simple model of the moving molecules in a gas. The gas exerts a pressure on the walls of its container because its molecules are continually bombarding the surface and rebounding from it.

Real and ideal gases

Most common gases are made up of molecules.

An *ideal gas* is one which exactly obeys Boyle's law.

Ideal gases do not exist. However, real gases approximate to ideal gas behaviour at low densities and at temperatures well above their liquefying points.

The ideal gas equation

From the equation of state, pV/T = constant. The constant can have different values depending on the type and mass of gas. However, if the amount of gas being considered is one mole (6.02×10^{23} molecules), then the constant is the same for all gases:

for one mole of any gas $\quad pV = RT$

R is called the **universal molar gas constant**.

Its value is 8.31 J mol^{-1} K^{-1}.

for n moles of any gas $\quad pV = nRT$

Note:
- The number of moles $n = m/M$, where m is the mass of the gas and M is its **molar mass** (the mass per mole).

The equation can also be given in a form for the number of molecules (or atoms if the gas is monatomic) by using a different constant – the **Boltzmann constant k**, $k = 1.38 \times 10^{-23}$ J K^{-1}

For N molecules of any gas

$pV = NkT$

Atoms and molecules

The Avogadro constant, $N_A = 6.02 \times 10^{23}$ mol^{-1}

One **mole** of any substance contains 6.02×10^{23} molecules

This number of molecules is useful because it is the number of molecules in 0.012 kg of pure carbon-12, that is one mole of pure carbon-12.

When calculating the mass of a mole of any particles you must take into account whether they are atoms, molecules or other particles. A mole of hydrogen atoms has a mass of 0.001 kg but a mole of hydrogen gas has a mass of 0.002 kg because the molecule contains 2 atoms.

Examples

Substance	Molar mass/kg mol^{-1}
Carbon dioxide (CO_2)	0.044
Hydrogen gas (H_2)	0.002
Helium gas (He)	0.004
Nitrogen gas (N_2)	0.028
Oxygen gas (O_2)	0.032

Kinetic energy and temperature

In an ideal gas the translational kinetic energy is equal to the internal energy. The Kelvin temperature scale is a thermodynamic scale, so the internal energy is directly proportional to the temperature in kelvin.

This means that the mean translational kinetic energy of a molecule in an ideal gas is directly proportional to the temperature in kelvin.

The mean translational kinetic energy of N molecules,

$$E = \frac{[\frac{1}{2} m c_1^2 + \frac{1}{2} m c_2^2 + \dots \frac{1}{2} m c_N^2]}{N}$$

$$E = \frac{1}{2} m \frac{[c_1^2 + c_2^2 + \dots c_N^2]}{N} = \frac{1}{2} m <c^2>$$

Where $<c^2>$ is the mean square speed of the molecules

$$<c^2> = \frac{[c_1^2 + c_2^2 + \dots c_N^2]}{N}$$

$$\frac{1}{2} m <c^2> \propto T$$

c_{rms} is called the root mean square speed, or the rms speed and is given by:

$$c_{rms} = \sqrt{<c^2>}$$

The mean translational kinetic energy of the molecules, E is given by:

$$E = \frac{3kT}{2}$$

Where k is the Boltzmann constant and T is the temperature in kelvin.

Example

In a sample of gas, the mass of each gas molecule is 3.3×10^{-27} kg. The temperature is 80° C, estimate the root mean square speed of the gas molecules.

$$\frac{1}{2} m <c^2> = \frac{3kT}{2}$$

$$<c^2> = \frac{3kT}{m}$$

$$c_{rms} = \sqrt{\frac{3 \times (1.38 \times 10^{-23} \text{ J K}^{-1}) \times (273 + 80) \text{ K}}{3.3 \times 10^{-27} \text{ kg}}}$$

$$= 2104 \text{ m s}^{-1}$$

Unit G485
Fields, particles and frontiers of physics

Module 1: Electric and magnetic fields

5.1.1 Electric fields

Electric force

Charges attract or repel each other with an **electric force**. If point charges Q and q, are a distance r apart, and F is the force on each, then according to **Coulomb's law**:

$$F \propto \frac{Qq}{r^2}$$

This is an example of an **inverse square law**. If r doubles, the force F drops to one quarter, and so on.

With a suitable constant, the above proportion can be turned into an equation:

$$F = \frac{kQq}{r^2}$$

The unit of charge for Q and q is the coulomb (C).

The value of k is found by experiment. It depends on the **medium** (material) between the charges. For a vacuum, k is 8.99×10^9 N m^2 C^{-2}, and is effectively the same for air.

In practice, it is more convenient to use another constant, ε_0, and rewrite the equation on the left in the following form:

In a vacuum $\qquad F = \dfrac{Qq}{4\pi\varepsilon_0 r^2}$

ε_0 is called the **permittivity of free space**. Its value is 8.85×10^{-12} C^2 N^{-1} m^{-2}.

Note:
- Although '4π' complicates the above equation, it simplifies others derived from it.
- In the above equation, if, say, Q and q are *like* charges (e.g. – and –), then F is *positive*. So a positive F is a force of *repulsion*. Similarly, it follows that a negative F is a force of *attraction*.
- The units of ε_0 can also be quoted as Fm^{-1}.

Example
Calculate the force on an electron in a hydrogen atom when it is is 5.3×10^{-11} m from the nucleus.

$$F = \frac{Qq}{4\pi\varepsilon_0 r^2}$$

$$F = \frac{(+1.6 \times 10^{-19}\,\text{C}) \times (-1.6 \times 10^{-19}\,\text{C})}{4\pi \times (8.85 \times 10^{-12}\,\text{C}^2\,\text{N}^{-1}\,\text{m}^{-2}) \times (5.3 \times 10^{-11}\,\text{m})^2}$$

$$= -8.2 \times 10^{-8}\,\text{N}$$

Electric field

Electric fields are created by electric charges.

> At a point in space, the electric field strength is the force per unit positive charge experienced by a point charge at that point in space.

If a charge feels an electric force, then it is in an **electric field**. If a charge q feels a force F, then the **electric field strength** E is:

$$E = \frac{\text{electric force}}{\text{charge}} \qquad \text{In symbols} \quad E = \frac{F}{q} \qquad (1)$$

For example, if a charge of 2.0 C feels an electric force of 10 N, then E is 5.0 N C^{-1}.

Note:
- Electric field strength is a vector. Its direction is that of the force on a positive (+) charge.

The force acting on a point charge in an electric field can be found by rearranging the equation above:

$$F = qE$$

electric field due to Q:

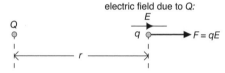

The point charge Q produces an electric field which acts on a small charge q. As $F = qE$ and

$$F = \frac{Qq}{4\pi\varepsilon_0 r^2} \qquad \text{it follows that} \quad E = \frac{Q}{4\pi\varepsilon_0 r^2} \qquad (2)$$

Example
By considering a helium nucleus as a point charge, what is the electric field 1 nm from the nucleus?

$$E = \frac{Q}{4\pi\varepsilon_0 r^2}$$

$$E = \frac{(2 \times 1.6 \times 10^{-19}\,\text{C})}{4\pi \times (8.85 \times 10^{-12}\,\text{C}^2\,\text{N}^{-1}\text{m}^{-2}) \times (1.0 \times 10^{-9}\,\text{m})^2}$$

$$= 2.88 \times 10^{-9}\,\text{N C}^{-1}$$

Electric field lines

Electric field lines are used to represent the electric field around charged objects.

- The direction of the field lines is the direction that a small stationary positive charge would move.
- The field is stronger where the lines are closer together.
- A uniform field has parallel, equally spaced field lines. The field between two oppositely charged parallel plates (but not near their edges) is uniform.
- The lines around point charges, and spherical charges, are radial and equally spaced. (Place your ruler right across the charge to draw these lines, but do not draw the part of the line inside the charge or charged conductor.)

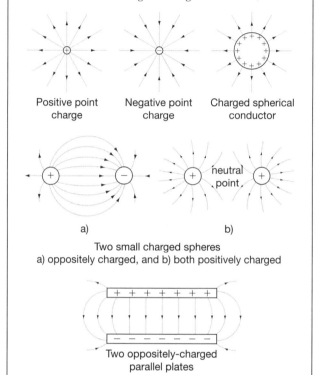

| Positive point charge | Negative point charge | Charged spherical conductor |

a)   b)

Two small charged spheres
a) oppositely charged, and b) both positively charged

Two oppositely-charged parallel plates

Electric field strength between parallel plates

The metal plates on the right have a small test charge q between them. The charge feels a force F. So, from equation (1),

$$E = \frac{F}{q}$$

There is a potential difference V between the plates. From the graph below, the potential gradient is $-V/d$.
So

$$E = \frac{V}{d}$$

The constant potential gradient means that the electric field is uniform.

The above equations show different aspects of electric field strength. If, say, E is $10\,N\,C^{-1}$, you can think of this either as a force of 10 N per coulomb or a potential drop of 10 V per metre.
(E can be expressed in $N\,C^{-1}$ or in $V\,m^{-1}$.)

Example

Two parallel plates are 2.0 cm apart and the potential difference between them is 1.2 kV. What is the electric field strength between the plates?

$$E = \frac{V}{d} = (1200\,V) \div (2.0 \times 10^{-2}\,m) = 60\,N\,C^{-1}$$

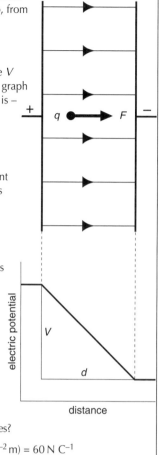

Comparing electric and gravitational fields

For particles of similar size, electric forces are very much stronger than gravitational ones. For example, electric forces hold atoms together to form solids.

Electric and gravitational fields have similar features. That is why the equations in this unit have a similar form to those on page 26. However, comparing equivalent equations, a minus sign may be present in one but absent from the other. This arises because of the differing force directions.

Gravity is always a force of attraction. Mass is always positive and it produces a gravitational field which is directed *towards* it.

Electric charges may attract or repel. However, if a charge is positive, then it produces an electric field which is directed *away* from it.

	Gravitational field	Electric field
Definition of field	Force on a unit mass	Force on a unit positive charge
Field	g	E
Equation for field strength	$g = F/m$	$E = F/q$
SI units of field strength	$N\,kg^{-1}$	$N\,C^{-1}$
Force acts on	All matter	Electrical charges
Direction of force	Always in direction of field	In direction of field for positive charge, opposite direction for negative charge.
Equation for force between two point objects	$F = -\dfrac{GmM}{r^2}$	$F = \pm\,\dfrac{Qq}{4\pi\varepsilon_0 r^2}$
Work done moving distance d	$W = mgd$	$W = Eqd$

Charged particles in a uniform electric field

In a uniform electric field, a positive charge will accelerate along the direction of the field lines and a negative charge will accelerate in the opposite direction to the field lines. If the charge is stationary, or already moving parallel to the field, the acceleration will speed it up or slow it down but will not cause any change of direction.

If a charge is already moving when the field is applied, then the acceleration will cause it to change direction. In a uniform field the particles will always follow a parabolic path:

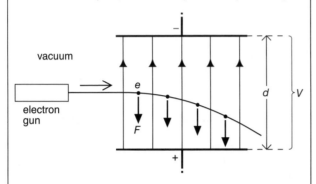

The electrons pass between two horizontal plates. The electric field strength between the plates is V/d.

force F on electron = electric field strength × charge

So
$$F = \frac{Ve}{d}$$

Note:
- The force on the electron does not depend on its speed.
- The force is always in line with the electric field.
- The path of the electrons is a *parabola* (the situation is similar to that of a projectile – where the gravitational field is at right angles to the horizontal motion).

5.1.2 Magnetic fields

Magnetic fields from currents

A current has a magnetic field around it. The greater the current, the stronger the field. The field round a current-carrying wire is shown on the right. The field direction is given by **Maxwell's screw rule**. Imagine a screw moving in a clockwise direction. The field turns the same way as the screw.

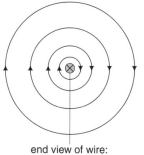

end view of wire:
current direction into paper

The field round a current-carrying coil (below) is similar to that round a bar magnet. The field direction can be worked out using either the screw rule or the **right-hand grip rule**. Imagine gripping the coil with your right hand so that your fingers curl the same way as the current direction. Your

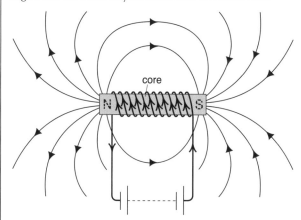

thumb is then pointing to the N pole.
The field is very much stronger if the coil has an iron **core**. Together, the coil and core form an **electromagnet**. With an iron core, the field vanishes when the current is turned off. However, a steel core keeps its magnetism. Permanent magnets are made using this principle.

Orbiting electrons in atoms are tiny currents. They are the source of the fields from magnets.

Magnetic force on a current

There is a force on a current in a magnetic field. Its direction is given by **Fleming's left-hand rule**:

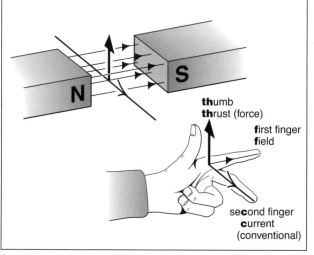

thumb
thrust (force)

first finger
field

second finger
current
(conventional)

Magnetic force and flux density

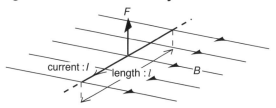

Above, a current-carrying wire is at right-angles to a uniform magnetic field. The field exerts a force on the wire. The direction of the force is given by Fleming's left-hand rule. The size of the force depends on the current I, the length l in the field, and the strength of the field. This effect can be used to define the magnetic field strength, known as the **magnetic flux density**, B:

$$F = BIl$$

B is a vector. The SI unit of B is the **tesla** (T). For example, if the magnetic flux density is 2.0 T, then the force on 2.0 m of wire carrying a current of 3.0 A is $2.0 \times 2.0 \times 3.0 = 12$ N.

If a wire is not at right angles to the field, then the above equation becomes

$$F = BIl\sin\theta$$

where θ is the angle between the field and the wire. As θ becomes less, the force becomes less. When the wire is *parallel* to the field, $\sin\theta = 0$, so the force is zero.

Magnetic force on a moving charge

Consider a length l of the wire at right angles to the magnetic field. It has a cross sectional area A, and the number density of the charges (the number per unit volume) is n, then the total number of charges in the section $N = nAl$:

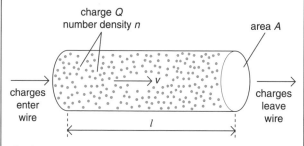

charge Q
number density n

area A

charges enter wire

charges leave wire

The force F_w on the current-carrying wire, due to the magnetic field on all the charges, is $F_w = BIl$

The force on a single charge in the wire $F = \dfrac{BIl}{nAl}$

This is the force on N charges, so the force, due to the magnetic field, on one charge is:

$$F = \frac{B\,n\,A\,Q\,v\,l}{nA}$$

The current in the wire is $I = nAQv$

$$F = BQv \qquad (1)$$

Note:
- As the speed increases, the force increases.
- The direction of the force is given by Fleming's left-hand rule.
- If the particle is travelling at an angle θ to the field, then the above equation becomes

$$F = BQv\sin\theta. \qquad (2)$$

Deflection of electrons by a magnetic field

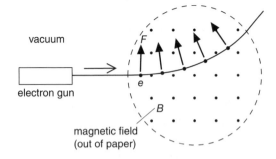

vacuum

electron gun

magnetic field
(out of paper)

Above, electrons travel at right angles to a uniform magnetic field. The force F on an electron is found by putting Q equal to e in equation (1)

$$F = Bev$$

Note:
- The force on the electron increases with speed.
- The force is always at right-angles to the direction of motion, as predicted by Fleming's left-hand rule. But in applying this rule, remember that the electron has negative charge, so electron motion to the *right* represents a conventional current to the *left*.
- The path of the beam is *circular*.

Mean what you say – say what you mean

For charged particles in magnetic fields, if the direction of the particle is towards or away from a magnetic pole (so that the field is in line with the motion of the particle), the force on the particle is zero. There is no deflection. This is shown by equation (2), when $\theta = 0$ or $180°$, $\sin \theta = 0$. It is incorrect to say that the charged particle is 'attracted' or 'repelled' by the magnetic field.

The maximum deflection occurs when the field and the particle motion are perpendicular to each other. Describing the directions clearly can be difficult. If you write the force is 'up' or 'down' do you mean towards the top or bottom of the paper, or do you mean into or out of the paper?

Make sure your answers are not ambiguous. Use diagrams if these help.

A beam of both positive and negative particles can be separated by a magnetic field. The force on the positive particle is in the opposite direction to the force on the negative particle, so the **deflection** will be in opposite directions initially, as shown below. It is incorrect to say that the particles will **move** in opposite directions.

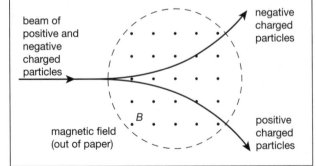

beam of positive and negative charged particles

negative charged particles

magnetic field
(out of paper)

positive charged particles

Circular motion in a magnetic field

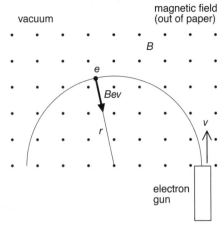

vacuum

magnetic field
(out of paper)

B

Bev

r

v

electron gun

Above, electrons leave the electron gun at speed v. They move in a circle because the magnetic force Bev supplies the necessary centripetal force (4.2.1 Circular motion page 22). So

$$Bev = \frac{m_e v^2}{r} \qquad \text{So} \qquad r = \frac{m_e v}{eB} \qquad (3)$$

Note:
- The radius of the circle is proportional to the speed.
- Increasing B decreases the radius.

Ion beams If atoms lose electrons, they become positive ions. These particles also have circular paths in a magnetic field. If Q is the charge on a particle, and m the mass, then equation (3) can be written in this more general form:

$$r = \frac{mv}{QB} \qquad (4)$$

Particles in electric and magnetic fields

An ion beam may contain ions with a range of speeds. Ions of only one speed can be selected by the method shown below. This uses the principle that the magnetic force on an ion depends on its speed, while the electric force does not:

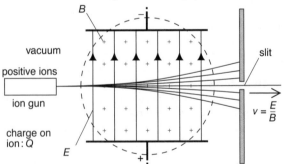

magnetic field into paper

B

vacuum

positive ions

ion gun

charge on ion: Q

E

slit

$v = \frac{E}{B}$

The magnetic field produces a force towards the top of the page, BQv. The electric field produces a force towards the bottom of the page, EQ. The only ions to pass through the slit are those for which these forces are equal. So, if $EQ = BQv$, the selected speed $v = E/B$. Faster ions are deflected upwards because BQv is more); slower ions are deflected downwards.

Mass spectrometer

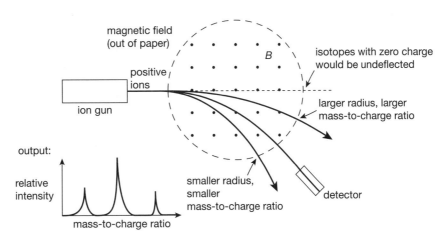

Equation (4) shows that the radius of curvature of the path of an ion in a magnetic field is $r = \dfrac{mv}{QB}$

To arrive at the detector the ions must travel in a path with the correct radius r. If the ions are travelling with the same speed v, then as B is increased, ions with a larger mass-to-charge ratio will arrive at the detector.

The output graphs or charts are arranged to have an x axis with the mass increasing linearly because this is the most useful format for the user who wants to measure the mass of the ions.

The output shows the relative intensity of the different isotopes detected, plotted against the mass-to-charge ratio for the ions. This ratio is the mass of the ion divided by the size of charge on the ion in units of e, the charge on the electron.

Usually all the ions have a charge of +1, because they have lost on electron, for example the mass-to charge ratio for an ion with mass m and charge +1 is $m \div 1 = m$. In this case the x axis is the same as the mass.

However, for an ion of mass $2m$ with a charge +2 (because it has lost two electrons), its mass-to-charge ration is $2m \div 2 = m$, so the peak will appear in the same place as the ion with mass m and charge +1, and this must be taken into account when analysing results.

The ratio of different isotopes is used in some dating techniques, e.g. carbon dating.

There are different designs of mass spectrometer and in some designs a speed selector is used that combines electric and magnetic fields, as described in 'Particles in electric and magnetic fields' to
produce a beam of ions all travelling at the same speed. This beam is then injected into a perpendicular magnetic field as shown above.

5.1.3 Electromagnetism

Magnetic flux

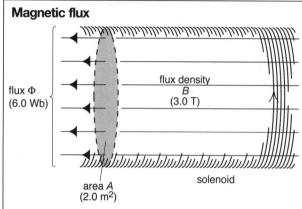

flux Φ
(6.0 Wb)

flux density
B
(3.0 T)

area A
(2.0 m²)

solenoid

Above, there is a uniform field of magnetic flux density B inside a solenoid of cross-sectional area A. The **magnetic flux** Φ is given by:

| flux = flux density × area | $\Phi = BA$ |

The SI unit of magnetic flux is the **weber** (**Wb**). For example, if B is 3 T and A is 2 m², then Φ is 6 Wb.

The definition of magnetic flux refers to a more general case, where the area, A, is not at right angles to the magnetic field, B.

The magnetic flux, Φ, through an area, A, is defined as the product of the magnetic flux density, B, and the projection of area A onto a surface at right angles to the flux.

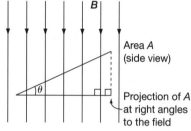

B

Area A
(side view)

θ

Projection of A
at right angles
to the field

Example

A metal disc of radius 0.5 cm is placed in a magnetic field of 0.03 T.

When the disc is perpendicular to the field $\theta = 0°$ $\cos\theta = 1$

$\Phi = (0.03\ T) \times \pi \times (0.5 \times 10^{-2}\ m)^2 \times 1 = 2.36 \times 10^{-6}$ Wb

When the disc is parallel to the field $\theta = 90°$ $\cos\theta = 0$

$\Phi = 0$

When the disc is at 30° to the field $\theta = 60°$ $\cos\theta = 0.5$

$\Phi = (0.03\ T) \times \pi \times (0.5 \times 10^{-2}\ m)^2 \times 0.5 = 1.18 \times 10^{-6}$ Wb

Note:
- Field lines do not really exist, but they can help you visualize what magnetic flux means. In the diagram above, each field line represents a flux of 1 Wb. There are 6 lines altogether, so the flux is 6 Wb. But there are 3 lines per m², so the flux density is 3 T.
- With flux, 'density' implies 'per m²' and not 'per m³'.
- 1 tesla = 1 weber per metre² i.e. 1 T = 1 Wb m⁻².

Magnetic flux linkage

When a coil of wire is in a magnetic field, the magnetic field passes through the area of each loop of wire, called a turn of the coil of wire

Magnetic flux linkage for a coil with N turns is equal to N multiplied by the magnetic flux through the coil.

Magnetic flux linkage = $N\ \Phi$

The units are weber turns (Wb turns)

Currents from magnetic fields

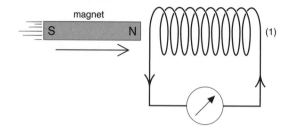

magnet

When one end of a bar magnet is moved into a coil, as above, an e.m.f. (voltage) is generated in the coil. This effect is called **electromagnetic induction**. The e.m.f. makes a charge flow in the circuit.

The induced e.m.f. (and the current) is increased when:
- the magnet is moved faster
- there are more turns on the coil
- a magnet giving a stronger field is used.

When the magnet is moved out of the coil, the e.m.f. is reversed. When the magnet is stationary, the e.m.f. is zero.

Varying a magnetic field can have the same effect as moving a magnet. Below, an e.m.f. is induced in the right-hand coil whenever the electromagnet is switched on or off. The current flows one way at switch-on and the opposite way at switch-off.

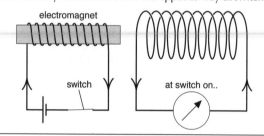

electromagnet

switch

at switch on..

Faraday's law of electromagnetic induction

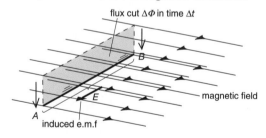

flux cut ΔΦ in time Δt

B

E

magnetic field

A

induced e.m.f

Above, a conductor AB is moving downwards at a steady speed through a magnetic field. It cuts through flux $\Delta\Phi$ in time Δt. As a result, an e.m.f. is induced between the ends of A and B of the conductor. According to **Faraday's law of electromagnetic induction**, the induced e.m.f. is proportional to the rate of cutting flux:

$$\text{induced e.m.f.} \propto \frac{\text{flux cut}}{\text{time taken}} \qquad \text{induced e.m.f.} \propto \frac{\Delta\Phi}{\Delta t}$$

With a constant, this can be turned into an equation. The constant is 1 because of the way the units are defined.

> Faraday's Law: In SI units, the magnitude of the induced e.m.f. is equal to the rate at which magnetic flux is cut.

$$\text{The magnitude of induced e.m.f.} = \frac{\text{flux cut}}{\text{time taken}} = \frac{\Phi}{t}$$

For a coil with N turns, the magnitude of induced e.m.f. is equal to the rate of change of flux linkage.

For example, if 6.0 Wb of flux are cut in 2.0 s, the induced e.m.f. is 3.0 V.

Lenz's law

> An induced e.m.f. is always in a direction that will oppose the change in flux which causes it.

For example, in the diagram above, the induced e.m.f.

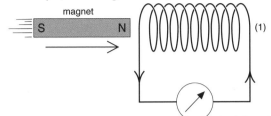

(1)

produces a current which causes a N pole at the left end of the coil so that the approaching magnet is repelled. If the magnet is moved the other way, the current direction reverses so that there is a pull on the magnet to oppose its motion.

Faraday's Law says that the induced e.m.f. is proportional (not equal) to the rate of change of flux. When the direction is taken into account, Lenz's law and Faraday's law together give the equation:

induced e.m.f. = – rate of change of magnetic flux linkage.

In the example of the magnet moving into the coil, work is done moving the magnet against a repulsive force. Lenz's law tells us that the direction of the e.m.f. opposes the change producing it and this means that work always has to be done to make the change. So Lenz's law follows from the law of conservation of energy. If the induced e.m.f was in the same direction as the change in flux which produced it, the magnet would be attracted into the coil, producing a larger e.m.f. which would attract it more and the amount of energy would increase, which is not possible.

Eddy currents When the aluminium disc on the right is spun between the magnetic poles, the e.m.f.s induced cause **eddy currents**. These set up a magnetic field which pulls on the poles and opposes the motion. So the disc quickly comes to a halt. Electromagnetic braking systems use this effect.

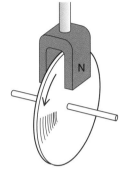

Generating alternating current (ac)

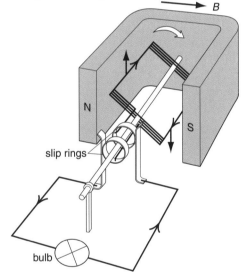

In the generator above, a coil is rotated in a magnetic field. This induces an e.m.f. in the coil, so a current flows.

The magnetic field, B, is horizontal from N to S. The coil is rotating so that the left hand side is moving up and the right hand side is moving down.

Fleming's right hand rule (first finger = field, second finger = current, thumb = motion) indicates that the current resulting from the induced e.m.f. will flow away from us on the left hand side and towards us on the right hand side.

When the coil reaches the vertical position, for a moment the flux linkage will not be changing so the induced e.m.f. will be zero. The side of the coil that was moving upwards will now be moving downwards (and the side that was moving downwards will now be moving upwards) so the induced e.m.f. and therefore the current will be in the opposite direction.

When the coil reaches the horizontal position the flux linkage will be changing most rapidly, so the induced e.m.f. and therefore the current will be a maximum.

When the current changes direction like this it is called alternating current (ac). A graph of the e.m.f. (or the current) will be a sine curve.

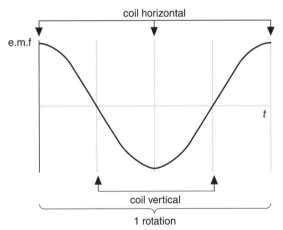

The size of the e.m.f. (and therefore the current) can be increased by:

- increasing the rate of change of flux linkage by turning the coil faster
- increasing the magnetic field by using a stronger magnet or a coil with a soft iron core
- increasing the flux linkage by increasing the number of turns on the coil.

Transformers

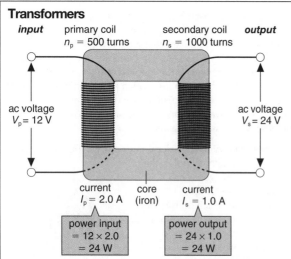

input — primary coil $n_p = 500$ turns
secondary coil $n_s = 1000$ turns — **output**

ac voltage $V_p = 12$ V
ac voltage $V_s = 24$ V

current $I_p = 2.0$ A   core (iron)   current $I_s = 1.0$ A

power input
$= 12 \times 2.0$
$= 24$ W

power output
$= 24 \times 1.0$
$= 24$ W

In a **transformer**, an alternating current in the **primary** (input) coil creates a changing magnetic flux in the core and secondary (output) coil which, induces an alternating e.m.f. in the **secondary** coil. A **step-up** transformer, as above, increases the voltage. A **step-down** transfomer reduces it.

Note that transformers will only work with an ac supply.

There is a link between the output and input voltages (V_s and V_p) and the numbers of turns (n_s and n_p) on the coils:

$$\frac{V_s}{V_p} = \frac{n_s}{n_p}$$

For the transformer in the diagram,

$$\frac{n_s}{n_p} = \frac{1000}{500} = 2.0 = \frac{24}{12} = \frac{V_s}{V_p}$$

If no power is wasted in the coils or core,

power output = power input

$$V_s I_s = V_p I_p$$

This means that a transformer which *increases* voltage will *decrease* current, and vice versa.

On a circuit diagram, a transformer symbol is:

n_p turns n_s

V_p core V_s

I_p I_s

input   output

In practice, transformers waste energy as heat, due to:
- resistance of the coils
- eddy currents produced by the changing flux (see page 49).

The core is laminated (layered) to reduce this.

For a practical transformer $V_s I_s = e V_p I_p$

where e is the efficiency (typically over 0.95).

Examples

An appliance designed to use a 12V supply uses a mains adapter containing a transformer.

The transformer will be a step-down transformer to transform the voltage from 230V to 12V.

It will have more turns on the primary than the secondary coil.

If there are 690 turns on the primary coil, then

$$\frac{V_s}{V_p} = \frac{n_s}{n_p}$$

$$n_s = \frac{(12 \text{ V}) \times (690)}{(230 \text{ V})} = 36 \text{ turns}$$

At a power station a step-up transformer is used to increase the voltage from 23.5 kV to the voltage for the transmission lines. There are 49 000 turns in the primary coil and 280000 turns in the secondary coil.

A step-up transformer to increase the voltage will have more turns on the secondary than the primary coil.

$$V_s = \frac{(280000) \times (23 \text{ kV})}{(49\,000)} = 130 \text{ kV}$$

Module 2: Capacitors and exponential decay

5.2.1 Capacitors

Capacitance

Capacitors store energy by the separation of charge (they are sometimes said to 'store charge' but this is not an accurate description).

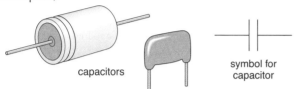

capacitors · symbol for capacitor

A capacitor can be charged by connecting a battery across it. The higher the p.d. V, the greater the charge Q stored. Experiments show that $Q \propto V$. Therefore, Q/V is a constant.

> The *capacitance* C of a capacitor is the charge Q per unit potential difference V.
>
> $$\text{capacitance} = \frac{\text{charge}}{\text{p.d.}} \qquad \text{In symbols } C = \frac{Q}{V}$$

The higher the capacitance, the more charge is stored for any given p.d.

Capacitance is measured in C V^{-1}, known as a *farad* (F). However, a farad is a very large unit, and the µF (10^{-6} F) is more commonly used for practical capacitors.

Example

The charge Q on a 2.0µF capacitor which has a p.d. of 9.0V across it is:

$Q = CV$

$Q = (2.0 \times 10^{-6} \text{ F}) \times (9.0\text{V}) = 18 \times 10^{-6} \text{ C}$

Capacitors in series

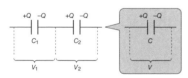

If C_1 and C_2 have a total capacitance of C, then C is the single capacitance which could replace them.

Two capacitors in series separate only the same charge Q as a single capacitor. So $V = Q/C$, $V_1 = Q/C_1$, and $V_2 = Q/C_2$.

But $V = V_1 + V_2$

So $\dfrac{Q}{C} = \dfrac{Q}{C_1} + \dfrac{Q}{C_2}$

or $\boxed{\dfrac{1}{C} = \dfrac{1}{C_1} + \dfrac{1}{C_2}}$

For example, if $C_1 = 3$ µF and $C_2 = 6$ µF, then

$1/C = 1/3 + 1/6 = 1/2$.

So $C = 2$ µF.

similarly $\dfrac{1}{C} = \dfrac{1}{C_1} + \dfrac{1}{C_2} + \dfrac{1}{C_3}$

Capacitors in parallel

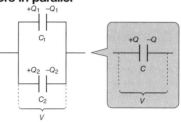

If C_1 and C_2 have a total capacitance of C, then C is the single capacitance which could replace them.

Capacitors in parallel each have the same p.d. across them. So $Q = CV$, $Q_1 = C_1V$, and $Q_2 = C_2V$.

Together, the capacitors act like a single capacitor with a larger plate area. So $Q = Q_1 + Q_2$

$$\therefore \qquad CV = C_1V + C_2V$$

$$\text{and} \qquad \boxed{C = C_1 + C_2}$$

For example, if $C_1 = 3$ µF and $C_2 = 6$ µF, then $C = 9$ µF.
Similarly $C = C_1 + C_2 + C_3$

Energy stored by a capacitor

Work must be done to charge up a capacitor. Electrical potential energy is stored as a result.

If a charge of 2.0 C is moved through a *steady* p.d. of 10 V, then,

work done $W = QV = 2.0 \text{ C} \times 10 \text{ V} = 20 \text{ J}$.

So the stored energy is 20 J. Numerically, this is the area under the graph below.

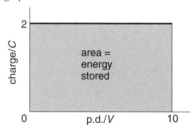

When a capacitor is being charged, Q and V are related as in the graph below. As before, the energy stored is numerically equal to the area under the graph, which is $\frac{1}{2}QV$.

> energy stored $= \frac{1}{2}QV$

As $C = \dfrac{Q}{V}$ this also gives $W = \frac{1}{2}CV^2$

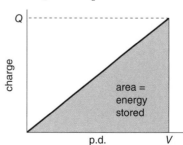

Circuits with capacitors
Example

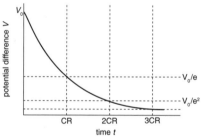

Calculate a) the total capacitance of the circuit, b) the charge on each capacitor, c) the p.d. across each capacitor.

a) The capacitance of the parallel combination of two 5.0 pF capacitors = 5.0 pF + 5.0 pF = 10.0 pF

Adding the capacitor in series, the total capacitance of the three capacitors is given by

$$\frac{1}{C} = \frac{1}{(4.0\,\text{pF})} + \frac{1}{(10.0\,\text{pF})}$$

$C = 1 \div 0.35 = 2.857$ pF   so the total capacitance = 2.9 pF

b) The total charge, the charge on the 4.0 pF capacitor and the charge on the 5.0 pF combination are all the same, so using $Q = CV$:

$Q = (2.857\,\text{pF}) \times (12\,\text{V}) = 34.28 \times 10^{-12}$ C so the charge on the 4.0 pF capacitor = 34 pC

The charge on each of the 5.0 pF capacitors will be the same, as the charge will be divided equally between them:
$Q = 34.28 \times 10^{-12}\,\text{C} \div 2 = 17.14 \times 10^{-12}\,\text{C}$

So the charge on each of the 4.0 pF capacitors = 17 pC

c) The p.d. across the 4.0 pF capacitor =
$$\frac{Q}{C} = \frac{(34.28 \times 10^{-12}\,\text{C})}{(4.0 \times 10^{-12}\,\text{F})} = 8.57\,\text{V}$$
So the p.d across the 4.0 pF capacitor = 8.6 V

The p.d across each of the 5.0 pF capacitors will be the same, as they are in parallel:
$$\frac{Q}{C} = \frac{(17.14 \times 10^{-12}\,\text{C})}{(5.0 \times 10^{-12}\,\text{F})} = 3.43\,\text{V}$$

The p.d across each of the 5.0 pF capacitors = 3.4 V

(check: 8.57 V + 3.43 V = 12 V)

Discharge of a capacitor

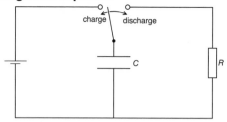

The capacitor above is charged from a battery and then discharged through a resistance R.

Graph A shows how, during discharge, the charge Q decreases with time t, according to the following equation:

$$Q = Q_0 e^{\frac{-t}{CR}}$$ where e = 2.718

CR is called the **time constant**. (It equals the time which the charge would take to fall to zero if the initial rate of loss of charge were maintained.) Increasing C or R gives a higher time constant, and therefore a slower discharge.

The gradient of the graph at any time t is equal to the current at that time.

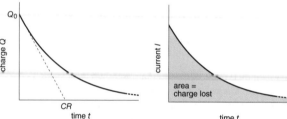

Graph A   **Graph B**

Graph B shows how the current decreases with time. The area under the graph is numerically equal to the charge lost.

The current, I, decreases according to the equation:
$I = I_0 e^{\frac{-t}{CR}}$

Graph C shows how the p.d. across the capacitor decreases with time.

The potential difference, V, across the capacitor also decreases with time according to the equation:

$$V = V_0 e^{\frac{-t}{CR}}$$

The current, charge and p.d. all decrease exponentially. The constant e is called the exponential number.

Notice that all three equations are of the form $x = x_0 e^{\frac{-t}{CR}}$

Exponential relationships

Many changes, such as the discharge of a capacitor or radioactive decay, are exponential changes.
The rate of change at any time is proportional to the amount at that time.

For exponential decay:

$$\frac{dx}{dt} = -kt \quad \text{or} \quad x = x_o e^{-kt}$$

A plot of x against time, t, gives a curve in which x changes by equal fractions for equal steps in t.

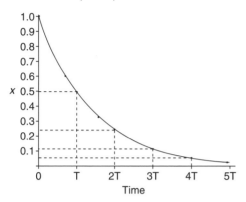

It is a property of the decay that for any size time interval chosen, the ratio between the amount at the start and the amount at the end of that size time interval is constant.

Example

The graph shows a time interval T in which x falls by 0.5.

The ratio of the amount at time T to the amount at time $2T = 0.5 : 0.25$ (ratio = 0.5).

The ratio of the amount at time $2T$ to the amount at time $3T = 0.25 : 0.125$ (ratio = 0.5).

The ratio of the amount at time $3T$ to the amount at time $4T = 0.125 : 0.0625$ (ratio = 0.5).

Notice too that for the fraction to fall from 0.6 to 0.3 (which is 0.5 x 0.6) the time is also T. Similarly from 0.8 to 0.4, or any other halving of the value of x.

It is difficult to see by eye if a curve is exponential. To test a relationship, either use log-linear graph paper or plot a graph of the natural logs against time.

Taking natural logs of the equation $Q = Q_0\, e^{\frac{-t}{CR}}$ gives

$$\ln Q = \ln Q_0 \; \frac{-t}{CR}$$

A graph of $\ln Q$ against t is a straight line with gradient $\frac{-1}{CR}$.

The diagram is a graph of $\ln (Q/C)$ against t/s.

Note: label the x-axis with the numerical value of the time t divided by the unit s. Label the y-axis with the natural log of the numerical value of the charge Q divided by the unit C.

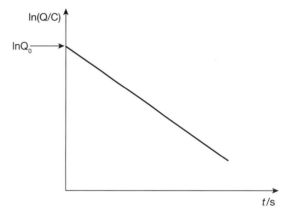

The gradient can be used to find the value of the time constant CR.

If you are given a graph like A (page 52) and asked to find CR, you can use the tangent at Q_0 as shown, but this may be difficult to draw. Another method is to use the time for Q to fall to 1/e of its original value Q_0.

$$\frac{Q_0}{e} = Q_0\, e^{\frac{-t}{CR}}$$

$$e^{-1} = e^{\frac{-t}{CR}}$$

$$-1 = \frac{-t}{CR}$$

$$t = CR$$

1/e is about 0.37 or 37%, but there is no need to remember this, as you can use any scientific calculator to calculate it when needed.

These methods can be used for radioactive decay graphs (see 5.3.3 Radioactivity page 61–63).

Uses of capacitors

Capacitors are used to store energy. When a capacitor is disconnected from a circuit it remains charged until it is discharged, so it can be used as a temporary battery.

Note:
- Although they are often said to store charge, the positive charge on one plate of the capacitor is exactly matched by the negative charge on the other plate. There is no net charge. When a charged capacitor is connected in a circuit charge will flow round the circuit from one plate to the other.

Flash photography

A camera usually has a low voltage battery, but the flash of light requires a large p.d. across a discharge tube. This is achieved by charging up a capacitor. The battery can only do this slowly, which is why there is often a noticeable delay before being able to use the flash again. Other components, including a transformer to step-up the charging voltage, are also required. The battery is not just connected directly to the capacitor as this would not result in a larger p.d. across it.

When the camera shutter is opened the capacitor is quickly discharged through the gas discharge tube and this ionises the gas producing a short burst of light.

Lasers for nuclear fusion

One method which scientists hope could be developed into a nuclear fusion power source, uses a number of very high power lasers focused on a pellet of fuel. The lasers vaporise the fuel, producing a plasma, in which nuclear fusion takes place.

Each laser uses flash-lamps to provide the initial light source which triggers the laser pulse – and the flash-lamps use capacitors.

For example, at the National Ignition Facility in the U.S.A. there are capacitor bays which store about 400 MJ of energy for each shot. The energy is stored for 60s and released in a 400 μs burst. The maximum current is more than 100 MA and the maximum power (for a fraction of a second) is more than 1TW. These capacitors provide the electrical energy for over 7000 flash-lamps, which provide the light energy for 192 lasers.

Back-up power supplies for computers

Computers, and, for example, telecommunications equipment and data centres, need an uninterruptible power supply to provide power immediately if the mains power supply fails. This is not the same as a standby generator which would take a few minutes to start-up. Capacitors are also used to supply the energy when batteries are being changed so that data is not lost.

Module 3: Nuclear physics

5.3.1 The nuclear atom

The alpha particle scattering experiment

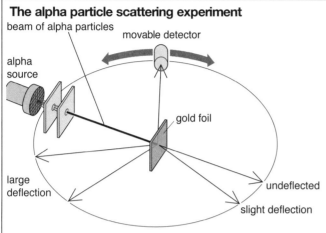

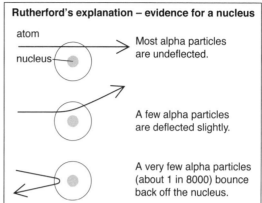

Rutherford's explanation – evidence for a nucleus

Most alpha particles are undeflected.

A few alpha particles are deflected slightly.

A very few alpha particles (about 1 in 8000) bounce back off the nucleus.

This experiment was carried out by Hans Geiger and Ernest Marsden under the supervision of Ernest Rutherford.

A thin piece of gold foil was bombarded with alpha particles, which are positively charged. Most of the alpha particles passed straight through the gold atoms. But a few were deflected slightly, and a very few (about 1 in 8000) were deflected back through very large angles.

These results came as a big surprise, and in 1911 Rutherford proposed a model of the atom, the '*Nuclear model*', to take account of them. According to this model, virtually all the mass of the atom is concentrated at the centre in a positively charged nucleus, with much lighter negatively charged electrons in orbit around it. Rutherford calculated the scattering this would cause and showed that the model of the atom was consistent with the experimental results.

Atoms

All matter is made from *atoms*. It would take more than a million million atoms to cover this full stop.

An atom has a tiny central *nucleus* made of *protons* and *neutrons* (apart from the simplest atom, hydrogen, whose nucleus is a single proton). Orbiting the nucleus are much lighter particles called *electrons*.

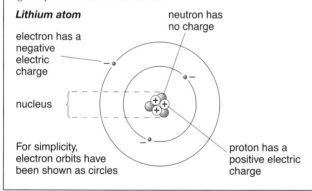

Lithium atom

electron has a negative electric charge

neutron has no charge

nucleus

For simplicity, electron orbits have been shown as circles

proton has a positive electric charge

An atom has the same number of electrons as protons, so the amounts of negative and positive charge balance.

Unlike charges (– and +) attract each other. This *electric force* holds electrons in orbit around the nucleus.

Like charges (– and –, also + and +) repel each other. However, the particles in the nucleus are held together by a *strong nuclear force*, which is strong enough to overcome the repulsion between the protons.

Atoms can stick together, in solids for example. The forces that bind them are attractions between opposite charges.

Moving electrons In metals, some of the electrons are only loosely held to their atoms. These *free electrons* can drift between the atoms. The electric current in a wire is a flow of free electrons.

If an atom gains or loses electrons, it is left with an overall – or + charge. Charged atoms are called *ions*.

Elements, nuclides, and isotopes

Everything is made from about 100 substances called **elements**. For most elements, a sample contains a number of different versions, called **isotopes**.

> **Isotopes** have the same number of protons (and electrons), but different numbers of neutrons.

This table shows some examples (italic numbers are for rarer isotopes).

Element	Electrons	Protons	Neutrons
hydrogen	1	1	0 or *1* or *2*
helium	2	2	*1* or 2
lithium	3	3	*3* or 4
carbon	6	6	6 or *7* or *8*
uranium	92	92	142 or *143* or 146

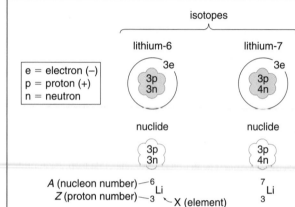

isotopes

lithium-6   lithium-7

e = electron (–)
p = proton (+)
n = neutron

nuclide   nuclide

A (nucleon number) —⁶Li
Z (proton number) —₃ ↘X (element)   ⁷Li
₃

Nuclide This is any particular version of a nucleus. Below left are simple models of the two naturally occurring nuclides of lithium, along with the symbols $^A_Z X$ used to represent them.

Nucleon number A As protons and neutrons are called **nucleons**, this is the total number of protons plus neutrons in the nucleus. It was once called the **mass number**. It is used when naming different isotopes – for example, carbon-12, carbon-13, carbon-14.

Proton number Z This is the number of protons in the nucleus (and therefore the number of electrons in a neutral atom). It was once called the **atomic number**.

Isotopes These are atoms with the same proton number but different nucleon numbers. They have the same electron arrangement and, therefore, the same chemical properties.

The following statements illustrate the meanings of the terms *element*, *nuclide*, and *isotope*.

- Lithium is an element.
- Lithium-6 is a nuclide; lithium-7 is a nuclide.
- Atoms of Lithium-6 and lithium-7 are isotopes.

Radioactive isotopes These have atoms with unstable nuclei. The nuclei break up, emitting **nuclear radiation**. The three main types of nuclear radiation are **alpha** particles, **beta** particles, and **gamma** rays (see 5.3.3 Radioactivity pages 61–63).

Atomic measurements

mass of proton	1.673×10^{-27} kg
mass of neutron	1.675×10^{-27} kg
mass of electron	9.110×10^{-31} kg
charge on proton	$+1.60 \times 10^{-19}$ C
charge on electron	-1.60×10^{-19} C
diameter of an atom	$\sim 10^{-10}$ m
diameter of a nucleus	$\sim 10^{-14}$ m

Atomic mass

The **unified atomic mass unit** (**u**) is used for measuring the masses of atomic particles. It is very close to the mass of one proton (or neutron). However, for practical reasons, it is defined as follows:

$$1 \text{ u} = \frac{\text{mass of carbon-12 atom}}{12}$$

Converting into kg, $1 \text{ u} = 1.66 \times 10^{-27}$ kg.

Note:
- The proton and neutron have approximately the same mass – about 1800 times that of the electron.
- $\sim$ means 'of the order of' i.e. 'within a factor ten of'.
- The diameter of an atom is $\sim 10^4$ times that of its nucleus. (Atom size varies from element to element.)
- Confusingly, the symbol e may be used to represent the charge on an electron (–) or a proton (+). Use e to mean the magnitude of the charge (1.6×10^{19} C) and write $+e$ for the charge on the proton and $-e$ for the charge on the electron.

Nuclear radii

Results of experiments show that the radius of a nucleus R is proportional to $A^{1/3}$ where A is the nucleon number.

Hence $R = R_0 A^{1/3}$

where R_0 is a constant having a value of 1.2×10^{-15} m.

Forces in the nucleus

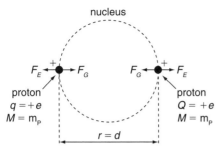

nucleus

Two protons in the nucleus repel each other because they are both positively charged.

Using **Coulomb's law** (5.1.1. Electric fields, page 42), the force of repulsion, F_E, between two protons, (charge, $e = 1.60 \times 10^{-19}$ C) at a separation of a small nuclear diameter ($d \sim 4.0 \times 10^{-15}$ m) is:

$$F_E = \frac{qQ}{4\pi\varepsilon_0 r^2} = \frac{(1.60 \times 10^{-19}\,\text{C})^2}{4\pi \times (8.85 \times 10^{-12}\,\text{C}^2\,\text{N}^{-1}\,\text{m}^{-2}) \times (4.0 \times 10^{-15}\,\text{m})^2}$$

$$= 14.4\,\text{N}$$

Protons have mass, so there is a gravitational force of attraction between two protons in the nucleus.

Using **Newton's Law of gravitation** (4.2.2 Gravitational fields page 26), the force of attraction, F_G, between two protons, (mass, $m_p = 1.673 \times 10^{-27}$ kg) at a separation of a small nuclear diameter ($d \sim 4.0 \times 10^{-15}$ m) is:

$$F_G = \frac{GMm}{r^2} = \frac{(6.67 \times 10^{-11}\,\text{N m}^2\,\text{kg}^{-2}) \times (1.673 \times 10^{-27}\,\text{kg})^2}{(4.0 \times 10^{-15})^2}$$

$$= 1.2 \times 10^{-35}\,\text{N}$$

So the gravitational attraction is negligible compared to the electrostatic repulsion. Another force must be holding the protons (and the neutrons) together in the nucleus. This force is called the **strong nuclear force**.

The strong nuclear force

In the nucleus, the nucleons (neutrons and protons) are bound together by the strong nuclear force. The strong force:
- is strong enough to overcome the Coulomb repulsion between protons, otherwise they would fly apart,
- has a short range, $\sim 10^{-15}$m, and does not extend beyond neighbouring nucleons,
- becomes a repulsion at very short range, otherwise the nucleus would collapse.

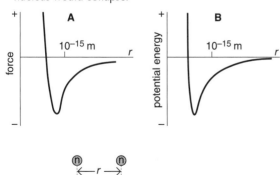

Graph A shows how the strong force varies with separation, for two neutrons. (An *attractive* force is *negative*.)

Graph B shows how the potential energy of the neutrons varies with separation. Minimum potential energy corresponds with the position of zero force in graph A.

Nuclear density

The mass of a nucleus	$= A \times$ the mass of a nucleon
The volume of a nucleus	$= \frac{4}{3}\pi(1.2 \times 10^{-15}A)^3$

The mass of a nucleon is 1.67×10^{-27} kg
The density of a nucleus is therefore 2.3×10^{17} kg m^{-3}

This is enormous when compared with the density of water (1000 kg m^{-3}). The difference is due to the fact that most of the volume of atoms is empty space. Only in neutron stars where the atoms are stripped of their electrons is it possible that such large densities exist on a large scale.

Nuclear decay reactions

Unstable nuclides are radioactive and eventually decay. When this happens the following are always conserved:

Momentum: the vector sum of the momentum of all the decay products is equal to the momentum of the parent nuclide.

Charge: the sum of the electric charge of all the products is equal to the charge of the parent nuclide.

Mass/energy: Energy is always conserved. In nuclear decay reactions the equivalence of mass and energy means that some mass is converted to energy, but the total of mass and energy remains constant. Energy is released as the kinetic energy of the products, and sometimes as a gamma ray photon.

In addition, when balancing nuclear equations, the sum of the nucleon numbers after the reaction is equal to the nucleon number of the parent nuclide.

The sum of the proton numbers after the reaction is equal to the proton number of the parent nuclide.

Alpha (α) decay

An alpha (α) particle consists of 2 protons and 2 neutrons, so it has a charge of $+2e$. It is identical to a nucleus of helium-4 and is represented by this symbol:

$$^{4}_{2}\text{He}$$

If an atom emits an α particle, its proton number is decreased, so it becomes the atom of a different element. For example, an atom of radium-226 emits an α particle to become an atom of radon-222, as shown by this equation:

$$^{226}_{88}\text{Ra} \quad \rightarrow \quad ^{222}_{86}\text{Rn} \quad + \quad ^{4}_{2}\text{He}$$

radium-226   radon-222   α particle

Note:
- Radon-222 and the α particle are the *decay products*.
- The nucleon numbers on both sides of the equation balance (226 = 222 + 4) because the total number of protons and neutrons is conserved (unchanged).
- The proton numbers balance (88 = 86 + 2) because the total amount of positive charge is conserved.
- Alpha decay tends to occur in heavy nuclides, because it produces a nuclide which is more stable.

Beta (β) decay

β⁻ decay This is the most common form of beta decay. The main emitted particle is an electron. It has a charge of $-1e$ and is represented by this symbol:

$$_{-1}^{0}e$$

Note:
- The beta particle is not a nucleon, so it is assigned a nucleon number of 0.
- The 'proton number' is –1 because the beta particle has an equal but opposite charge to that of a proton.

During β⁻ decay, a neutron is converted into a proton, an electron, and an almost undetectable particle with no charge and near-zero mass called an electron antineutrino (these particles are called electron neutrinos and antineutrinos because there are neutrinos associated with other particles). The electron is emitted, along with the antineutrino. For example, an atom of boron-12 emits a β⁻ particle to become an atom of carbon-12, as described by this equation:

$$_{5}^{12}B \quad \rightarrow \quad _{6}^{12}C \quad + \quad _{-1}^{0}e \quad + \quad \bar{\nu}$$

boron-12   carbon-12   β⁻ particle   antineutrino

Note:
- The nucleon numbers balance on both sides of the equation. So do the proton numbers.

β⁺ decay Here, the main emitted particle is a *positron*, with the same mass as an electron, but a charge of $+1e$. It is the *antiparticle* of an electron. For example, an atom of nitrogen-12 emits a β⁺ particle to become an atom of carbon-12, as described by the following equation:

$$_{7}^{12}N \quad \rightarrow \quad _{6}^{12}C \quad + \quad _{+1}^{0}e \quad + \quad \nu$$

boron-12   carbon-12   β⁺ particle   neutrino

Other types of decay

Electron capture

A proton in the nucleus captures an orbiting electron and becomes a neutron. The energy lost by the electron is emitted as an X-ray.

Transmutation

One element changes into another. This can occur when atoms are bombarded by other particles. For example, if a high-energy α particle strikes and is absorbed by a nucleus of nitrogen-14, the new nucleus immediately decays to form a nucleus of oxygen-17 and a proton. This is an example of a *nuclear reaction*. It can be described by the following equation:

$$_{7}^{14}N \quad + \quad _{2}^{4}He \quad \rightarrow \quad _{8}^{17}O \quad + \quad _{1}^{1}p$$

nitrogen-14   α particle   oxygen-17   proton

Note:
- The nucleon numbers balance on both sides of the equation. So do the proton numbers.

5.3.2 Fundamental particles

Fundamental particles

Fundamental particles are particles that cannot be divided into smaller particles.

The electron is a fundamental particle, one of a group of fundamental particles called **leptons**. The neutron and the proton are each made of three **quarks**, so they are not fundamental particles. They belong to the group of particles called **hadrons**.

In the ***standard model of particle physics*** (our model today), there are 12 fundamental particles from which matter is made (there will also be 12 corresponding antimatter particles). There are 6 types of *lepton* and 6 types of *quark* (and also 6 corresponding *antileptons* and 6 *antiquarks*). The main difference between quarks and leptons is that quarks feel the strong force and leptons do not.

Leptons

Leptons have no size and, in most cases, low or no mass. There are three generations of leptons, but only the first (the electron and its neutrino) occurs in ordinary matter. Muons and muon neutrinos are produced in the upper atmosphere by cosmic rays, but the tau has so far only been seen in laboratory experiments. The neutrino, ν, produced by beta$^-$ decay is the electron-antineutrino ν_e.

	Leptons	
Charge	^-e	**0**
1st generation	electron e^-	electron-neutrino ν_e
2nd generation	muon μ^-	muon-neutrino ν_μ
3rd generation	tau τ^-	tau-neutrino ν_τ

Quarks

Symmetry theory predicts that there should be 3 generations of quarks to match the 3 generations of leptons. They have a fractional charge of $+\frac{2}{3}e$ or $-\frac{1}{3}e$. The top quark is the most massive fundamental particle (almost 200 times the mass of the proton).

	quarks	
Charge	$+\frac{2}{3}e$	$-\frac{1}{3}e$
1st generation	up u	down d
2nd generation	charmed c	strange s
3rd generation	top t	bottom b

Note:
- Ordinary matter contains only the first generation of quarks. Very high energies are needed to make hadrons of other quark generations. These hadrons quickly decay into first generation particles.
- Individual quarks have never been detected.

Hadrons

Hadrons are particles made from quarks that are held together by the strong force. This force is so strong that quarks have never been found individually. There are two types of hadron:

Baryons are made of 3 quarks.

Mesons are made of a quark and an antiquark.

To describe how quarks can join together, each quark is assigned a baryon number of $\frac{1}{3}$ (each antiquark has ***baryon number*** $-\frac{1}{3}$).

All baryons have a baryon number of 1, antibaryons are −1, and all other particles, including mesons, are 0.

Similarly, the strange quark is assigned a strangeness of −1 (and the antistrange has +1) – all other quarks have strangeness 0. Strange particles have a long lifetime.

	quarks	charge	Baryon number	Strange ness	Name of particle	symbol
Baryon	uud	+1	+1	0	proton	p
	udd	0	+1	0	neutron	n
	$\overline{uud}$	−β1	−1	0	antiproton	$\overline{P}$
	$\overline{udd}$	0	−1	0	antineutron	$\overline{n}$
				0	↓(more baryons)	
Meson	$\overline{uu}$	0	0	0	pion (pi zero)	π^0
	$\overline{dd}$	0	0	0	pion (pi zero)	π^0
	$\overline{ud}$	+1	0	0	pion (pi plus)	π^+
	$\overline{du}$	−1	0	0	pion (pi minus)	π^-
	$\overline{ds}$	0	0	−1	kaon	k^0
	$\overline{ds}$	0	0	+1	kaon	k^0
	$\overline{us}$	+1	0	−1	kaon	k^+
	$\overline{su}$	−1	0	+1	kaon	k^-
		0			↓more mesons	

Note:
- π^0 has a very short lifetime, as the quarks will annihilate each other. The π^+ is the antiparticle of the π^- (so the symbol with the bar is not usually used). There is a similar pattern for other mesons.

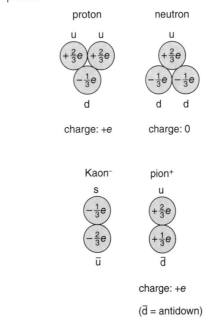

proton — charge: +e

neutron — charge: 0

Kaon$^-$

pion$^+$ — charge: +e

($\overline{d}$ = antidown)

Conservation laws

There are conservation laws for *momentum* and *mass energy*. As mass and energy are equivalent, the total energy must include the rest energy (see 5.3.4 Nuclear fission and fusion, page 64). Particles have various **quantum numbers** assigned to them. These are needed to represent other quantities which may be conserved during interactions. For example:

Charge In any interaction, this is conserved: it balances on both sides of the equation.

Lepton number This is +1 for a lepton, −1 for an antilepton, and 0 for any other particle. For example, a 'free' neutron decays, after about 15 minutes, like this:

$$\text{neutron} \rightarrow \text{proton} + \text{electron} + \text{antineutrino}$$
$$(0) \qquad (0) \qquad (+1) \qquad (-1)$$

The numbers (in brackets) have the same total, 0, on both sides of the equation, so lepton number is conserved. This applies in any type of interaction.

Baryon number This is +1 for a baryon, −1 for an antibaryon, and 0 for any other particle. It is conserved in all interactions.

Strangeness This is needed to account for the particular combinations of 'strange particles' (certain hadrons) produced in some collisons. It is conserved in strong and electromagnetic interactions, but not in all weak ones.

Charm relates to the likelihood of certain hadron decays.
Spin relates to a particle's angular momentum.
Topness and **bottomness** are further quantum numbers.

Matter and antimatter

Most matter particles, such as the proton, electron, and neutron, have corresponding **antiparticles**. These have the same rest mass as the particles but opposite charge and spin.

Apart from the antiparticle of the electron e⁻, which is the positron e⁺, antiparticles are given the same symbol as the particle but with a bar over the top.

Particle				Antiparticle	
e	electron	−	+	positron (antielectron)	e⁻
p	proton	+	−	antiproton	p̄
v	neutrino	spin		antineutrino	$\bar{v}$

When a particle and its antiparticle meet, in most cases, they **annihilate** each other and their mass is converted into energy as given by $\Delta E = \Delta mc^2$ (see page 64). For example, the annihilation of an electron and positron may produce a pair of gamma photons.

Note:
• There are far more particles than antiparticles in the Universe, so annihilation is extremely rare.

Other interactions

Energy can also be converted into mass. For example, if a gamma photon has at least 1.02 MeV of energy, it may, when passing close to a nucleus, convert into an electron-positron pair (total rest mass, or energy = 1.02 MeV see page 64). In high-energy collisions, heavier particles (and antiparticles) may materialize from the energy supplied.

The weak force and beta decay

The proton is stable, but all other hadrons outside the nucleus are unstable. Eventually they will decay, including the neutron. Inside the nucleus, whether protons or neutrons decay depends on whether they are needed for stability. If the nucleus is unstable, then the weak force causes one of the beta decays to occur.

Particle equation:

$$^{1}_{0}n \rightarrow ^{1}_{1}p + ^{0}_{-1}e + \bar{v}$$

in terms of quarks:

$$udd \rightarrow uud + e^- + \bar{v}$$
$$d \rightarrow u + e^- + \bar{v}$$

Particle equation:

$$p \rightarrow n + e^+ + v$$

Fundamental forces

Force	Range/m	Relative strength	Effects, e.g.
strong	~10⁻¹⁵	1	Holding nucleons in nucleus
electro-magnetic	∞	~10⁻²	Holding electrons in atoms; holding atoms together
weak	~10⁻¹⁷	~10⁻⁶	ß decay; decay of unstable hadrons
gravitational	∞	~10⁻³⁹	Holding matter in planets, stars, and galaxies

Particles *interact* by exerting forces on each other. There are four known types of force in the Universe (see chart above). As electric and magnetic forces are closely related, they are regarded as different varieties of one force, the electromagnetic. **Grand unified theories (GUTs)** seek to link the strong, weak, and electromagnetic forces. Gravitational force has yet to be linked with the others. It is insignificant on an atomic scale.

in terms of quarks:

$$uud \rightarrow udd + e^+ + v$$
$$u \rightarrow d + e^+ + v$$

When there is a nuclear reaction, the conservation of momentum and energy can be used to calculate the energy and momentum of the decay products. In beta decay, the beta particles are emitted with a range of possible energies. This is how the **neutrino** was discovered. There has to be another particle with the missing energy and momentum, otherwise the conservation laws are violated.

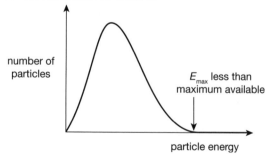

Neutrinos have very small rest mass, so small that it was initially thought to be zero.

5.3.3 Radioactivity

Radioactive decay

Apart from the hydrogen nucleus, which is just one proton, all nuclei are combinations of protons and neutrons. Some of these combinations of protons and neutrons are stable. Others are unstable and spontaneously disintegrate.

The stable light nuclei have roughly equal numbers of protons and neutrons, whereas stable heavier nuclei have about 50% more neutrons than protons.

All nuclei with more than 83 protons are unstable.

The disintegration is called **radioactive decay** and usually involves the emission of alpha particles or beta particles, and sometimes gamma rays, (see 5.3.1 The nuclear atom, pages 55–58). Radioactive decay is random and unpredictable. It is not possible to tell when a particular nucleus will decay, nor is it possible to speed up or slow down the process by any physical or chemical method.

Properties of alpha particles, beta particles, and gamma rays

Type	α	β	γ
nature	$2p + 2n$	e	electromagnetic photon
charge	$+2e$	$-1e$	no charge
speed (typical) (c = speed of light)	$0.1c$	up to $0.9c$	c
energy (typical)	10 MeV	0.03 to 3 MeV	1 MeV
ionizing effect: ion pairs per mm in air	$\sim10^5$	$\sim10^3$	~1
penetration (typical)	stopped by: 30-50 mm air a sheet of paper	stopped by: 3-5 mm aluminium	intensity halved by 100 mm lead
effect of magnetic field (B out of paper) *not to scale*		slow / fast	(undeflected)

Note:
- α, β, and γ emissions all cause *ionisation* – they remove electrons from atoms (or molecules) in their path. The removed electron (–) and the charged atom or molecule (+) remaining are called an *ion pair*. The ionised material can conduct electricity.
- α particles interact the most with atoms in their path, so they are the most ionising and the least penetrating.

- Unlike α particles, β particles are emitted from their source at a range of speeds.
- For γ radiation emitted from a point source in air, intensity $\propto$ 1/(distance from source)^2.
- γ radiation is not stopped by an absorber, but its intensity is reduced.
- α and β particles are deflected by magnetic fields, as predicted by Fleming's left-hand rule (see page 45). They are also deflected by electric fields.

Activity

The **activity** of a radioactive source is the number of disintegrations occurring within it per unit time.

The SI unit of activity is the *becquerel* (**Bq**):

1 becquerel = 1 disintegration s^{-1} (1 Bq = 1 s^{-1})

The activity of a typical laboratory source is ~ 10^4 Bq.

Each disintegration produces an α or β particle and, in many cases, γ radiation as well. The γ radiation is emitted as a *photon*. Particles and photons cause pulses in a detector, so the count rate is a measure of the activity of the source.

Note:
- The activity of a source is unaffected by chemical changes or physical conditions such as temperature. However, it does decrease with time.

Smoke detectors These contain a tiny α source which ionizes the air in a small chamber so that it conducts a current. Smoke particles entering the chamber attract ions and reduce the current. This is sensed by a circuit which triggers the alarm.

A smoke detector may use americium-241, which has a half-life of 432 years and is an α-emitter. The particles have short range – from the source to the detector – and are stopped by smoke. The activity does not change very much over a long period of time, so the rate of α emission will stay fairly constant, and the smoke detector should continue to work for years.

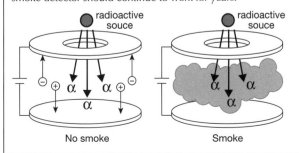

The decay law

Unstable nuclei disintegrate spontaneously and at random. However, the more undecayed nuclei there are, the more frequently disintegrations are likely to occur. For any particular radioactive nuclide, on average

activity ∝ number of undecayed nuclei

If N is the number of undecayed atoms after time t, then the activity A, is:

$$A = \lambda N \tag{1}$$

λ is called the **radioactive decay constant**. Each radioactive nuclide has its own characteristic value. (Note that the symbol λ is also used for wavelength.)

The decay constant is defined by equation (1).

Because the rate of decay (A) is proportional to the amount present (N) at any time, radioactive decay is an exponential decay.

The activity, A, drops during time, t, as given by the equation:

$$A = A_0\, e^{-\lambda t} \tag{2}$$

Where A_0 is the initial activity at the start, and e is the exponential constant , e = 2.718.

Alternatively, the exponential decay can be written in terms of the number of radioactive nuclei remaining , N, where N_0 is the initial number of undecayed nuclei at the start:

$$N = N_0 e^{-\lambda t} \tag{3}$$

A graph of A against t has the form shown below. The graph is an *exponential* curve.

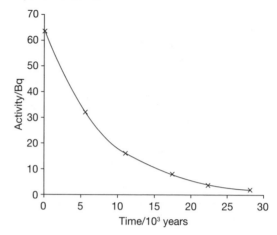

A graph of N against t will have a similar, exponential, shape.

More about the decay law equation

In calculus notation, because activity, A, is a rate of change of the number of nuclei present, N, with time, t, and because it is a decrease:

$$A = -\frac{dN}{dt}$$

so equation (1) can be written $\dfrac{dN}{dt} = -\lambda N$

Half-life

There are two alternative definitions for this term.

The **half-life** of a radioactive nuclide is:

A the average time taken for the number of undecayed nuclei to halve in value,

B the average time taken for the activity to halve.

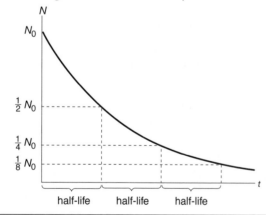

Version A is illustrated in the graph (left). One feature of the exponential curve is that the half-life is the same from whichever point you start.

In equation (3), the half life, $t_{\frac{1}{2}}$, is the value of t for which $N = N_0/2$.

Substituting this in the equation, taking natural logs, and rearranging gives

$$t_{\frac{1}{2}} = \frac{0.693}{\lambda} \qquad (0.693 = \ln 2)$$

Half-lives for some nuclides	
potassium-40	1.3×10^9 years
plutonium-239	24 400 years
carbon-14	5730 years
strontium-90	28 years
magnesium-28	21 hours
radon-224	55 seconds

Examples of exponential decay

Decay of radioactive nuclei

The rate of decrease of the number of radioactive nuclei is proportional to the number of radioactive nuclei remaining to decay.

Decay of charge on a capacitor

The rate that charge leaves a capacitor is proportional to the amount of charge remaining to be transferred.

Relationships where the rate of change of something depends on the unchanged amount remaining are exponential decays. They are described by equations of the form:

$y = y_0 e^{-kx}$ where the initial value of $y = y_0$ (when $x = 0$) and k is called the decay constant.

For radioactive decay, where N is the number of radioactive nuclei present:

$N = N_0 e^{-\lambda t}$

The decay constant, $k = \lambda$, the radioactive decay constant.

(There is also the exponential equation $A = A_0 e^{-\lambda t}$ where A is the activity.)

For the decay of charge on a capacitor, where Q is the charge, C is the capacitance and R is the resistance of the circuit:

$Q = Q_0 e^{\frac{-t}{CR}}$

The decay constant $k, = \dfrac{1}{CR}$, so, unlike radioactive decay,

instead of a special capacitor decay constant, the value quoted for a circuit is the time constant = CR.

(There are also the exponential equations $V = V_0 e^{\frac{-t}{CR}}$ and $I = I_0 e^{-\lambda t}$ where V is the voltage across the capacitor and I is the current in the circuit.)

To find the value of the decay constant

It is very difficult to judge whether a graph you have plotted by hand is exponential of some other curve.

To test whether a relationship is exponential, or to find the radioactive decay constant or time constant. It is best to use natural logs. Because taking logs of the equations gives, for example

$\ln(N) = \ln(N_0) - \lambda t \ln(e)$ where $\ln(e) = 1$

$\ln(N) = \ln(N_0) - \lambda t$

This is the equation of a straight line. A graph of $\ln(N)$ against t has a gradient $-\lambda$

or

$\ln(Q) = \ln(Q_0) - \dfrac{t}{CR} \ln(e)$ where $\ln(e) = 1$

$\ln(Q) = \ln(Q_0) - \dfrac{t}{CR}$

This is the equation of a straight line. A graph of $\ln(Q)$ against

t has a gradient $\dfrac{-1}{CR}$

(If the graph is not a straight line then the relationship is not exponential.)

Example

Finding the time constant for a discharging RC circuit by measuring the voltage across the capacitor

Time t/s	Voltage V/V	ln(V/V)
0	4.1	1.40
10	3.5	1.25
20	3.0	1.10
30	2.6	0.96
40	2.3	0.83
50	1.9	0.64
60	1.7	0.53
70	1.5	0.41
80	1.2	0.18
90	1.1	0.10

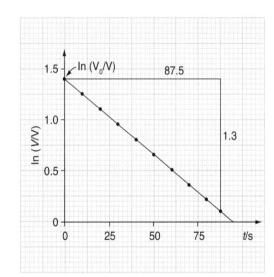

gradient $= \dfrac{-1.3}{87.5} = -0.0149$

gradient $= -\dfrac{1}{CR}$ where CR = time constant

so time constant $CR = 1 \div 0.0149 = 67$ s

5.3.4 Nuclear fission and fusion

Energy and mass

One conclusion from Einstein's **theory of relativity** is that energy and mass are equivalent. If an object gains energy, it gains mass. If it loses energy, it loses mass. The change of energy ΔE is linked to the change of mass Δm by this equation:

$$\Delta E = \Delta mc^2$$

where c is the speed of light: 3.0×10^8 m s^{-1}

c^2 is so high that energy gained or lost by everyday objects produces no detectable change in their mass. However, the energy changes in nuclear reactions produce mass changes which are measurable. For example, when a fast α particle is stopped, its mass decreases by about 0.2%. The mass of an object when it is at rest is called its **rest mass**.

Mass defect and binding energy

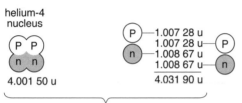

helium-4 nucleus

4.001 50 u   4.031 90 u

difference = mass defect = 0.030 40 u

A helium-4 nucleus is made up of 4 nucleons (2 protons and 2 neutrons). The calculation above shows that the nucleus has less mass than its four nucleons would have as free particles. The nucleus has a **mass defect** of 0.030 40 u.

The reason for the mass defect is as follows. In the nucleus, the nucleons are bound together by a strong nuclear force. As work must be done to separate them, they must have less potential energy when bound than they would have as free particles. Therefore, the separate particles have more energy than when they are combined in the nucleus, and this energy is called the **binding energy**.

> The nuclear **binding energy** is the energy needed to separate a nucleus into its individual protons and neutrons.

(The binding energy of an atom also includes the energy required to separate the electrons, but this is very small when compared to the energy required to separate the nucleons.)

The binding energy of the helium-4 nucleus
= mass defect $\times$ 931 MeV u^{-1}
= 0.03040 $\times$ 931 MeV = 28.3 MeV.

Note:

- Take care not to confuse the meaning of 'binding energy'. It is the energy needed to *separate* the nucleons – *not* the energy needed to bind them together.

Example

Iron has 26 protons. The binding energy of the iron-56 nucleus is $\Delta E = \Delta m c^2$

[proton rest mass , $m_p = 1.673 \times 10^{-27}$ kg, neutron rest mass, $m_n = 1.675 \times 10^{-27}$ kg, speed of light, $c = 3.0 \times 10^8$ ms^{-1}]

mass defect $\Delta m = (26\, m_p + 30\, m_n)$

$\Delta E = (26 \times 1.673 \times 10^{-27}$ kg $+ 30 \times 1.675 \times 10^{-27}$ kg$) \times (3 \times 10^8$ ms$^{-1})^2$

$\Delta E = 8.4 \times 10^{-9}$ J

This can also be expressed as units of MeV:
$\Delta E = 8.4 \times 10^{-9} \div 1.6 \times 10^{-13}$ MeV = 52750 MeV

With nuclear particles, energy is often measured in MeV (the electronvolt, eV, is defined in AS Physics on page 67):

$$1 \text{ MeV} = 1.60 \times 10^{-13} \text{ J}$$

From data on mass changes, scientists can calculate the energy changes taking place. With nuclear particles, mass is usually measured in u. By converting 1 u into kg and applying $\Delta E = \Delta mc^2$, it is possible to show that

| 1 u change in mass | is equivalent to | 931 MeV change in energy |

Binding energy per nucleon

> The **binding energy per nucleon** of the nucleus is the energy needed to separate a nucleus into its individual protons and neutrons, divided by the number of nucleons.

Although larger nuclei have a greater binding energy, the binding energy per nucleon is not constant.

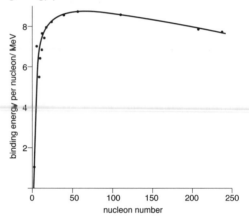

The stability of a nucleus depends on the **binding energy per nucleon**. The graph above shows how this varies with nucleon number. The line gives the general trend; points for some individual nuclides have also been included.

Note:

- Nuclei near the 'hump' of the graph are the most stable, because they have most binding energy per nucleon. They need the most energy to separate into their constituent nucleon.
- If nucleons become rearranged so that they have a *higher* binding energy per nucleon, there is an *output* of energy.

Radioactive decay Unstable nuclei decay to form more stable products, so energy is released. In α decay, for example, this is mostly as the kinetic energy of an α particle. When the α particle collides with atoms, it loses E_k and they speed up. So radioactive decay produces heat.

Nuclear reactions The **fission** and **fusion** reactions on the next page give out energy. During fission, heavy nuclei split to form nuclei nearer the 'hump' of the graph. During fusion, light, nuclei *fuse* (join) to form heavier ones.

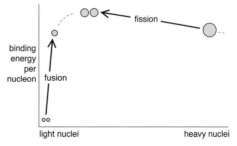

Nuclear fission

During **nuclear fission**, a heavy nucleus (e.g. of uranium or plutonium) splits to form two nuclei of roughly the same mass, plus several neutrons. Rarely, fission happens spontaneously. More usually, it occurs when a neutron hits and is captured by the nucleus. For example, here is a typical fission reaction for uranium-235:

$$^{235}_{92}U + ^{1}_{0}n \rightarrow ^{144}_{56}Ba + ^{90}_{36}Kr + 2^{1}_{0}n$$

The reaction releases energy, mostly as E_k of the heavier decay products. So fission is a source of heat.

Note:
- The energy released per atom by fission (about 200 MeV) is about 50 million times greater than that per atom from a chemical reaction such as burning.

Chain reaction The fission reaction above is started by one neutron. It gives off neutrons which may cause further fission and so on in a **chain reaction** (see diagram at right):

Uncontrolled chain reactions are used in nuclear weapons. Controlled chain reactions take place in **nuclear reactors** (see right) and release energy at a steady rate. The most commonly used fissionable material is uranium-235.

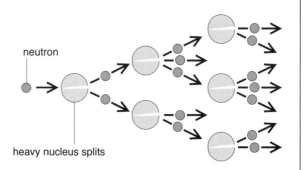

neutron

heavy nucleus splits

To maintain a chain reaction, a minimum of one neutron from each fission must cause further fission. However, to achieve this, these problems must be overcome:
- If the fission material is less than a certain **critical mass**, too many neutrons escape without hitting nuclei.
- The fission of uranium-235 produces medium-speed neutrons. But slow neutrons are better at causing fission.
- Less than 1% of natural uranium is uranium-235. Over 99% is uranium-238, which absorbs medium-speed neutrons without fission taking place.

Thermal reactors

In a nuclear power station, the heat source is usually a **thermal reactor**. (Otherwise, the layout is as for a fuel-burning station where steam turns the turbines.) In the reactor, there is a steady release of heat as fission of uranium-235 takes place. It is known as a *thermal* reactor because the neutrons are slowed to speeds associated with thermal motion.

Nuclear fuel This is uranium dioxide in which the natural uranium has been enriched with extra uranium-235. 1 kg of this fuel gives as much energy as about 25 tonnes of coal.

Moderator This is a material which slows down the medium-speed neutrons produced by fission. Some reactors use graphite as a moderator. The PWR uses water as a moderator.

Control rods These are raised or lowered to control the rate of fission. They contain boron, which absorbs neutrons.

Coolant (e.g. water or carbon dioxide gas) This carries heat from the reactor to the heat exchanger. In the PWR, this is pressurized water.

☐ pressurized water, moderator, and coolant   ☐ steam   ■ water

Containment structure   steam line

Fuel & control rods

steam   heat exchanger

turbine and generator

pump

condenser cooling water

pump

reactor   standard power station design

Pressurised water reactor (PWR)

Peaceful uses of nuclear fission

Generating electricity in thermal reactors

The heat from nuclear fission can be used as the power source for generating electricity in thermal power stations, as described above. There are benefits and risks:

Benefits of thermal reactors include:
- high energy output for the amount of fuel used
- no carbon dioxide produced in the reactor (although some may be produced in mining and preparing the fuel)
- many useful isotopes made by bombarding stable isotopes with neutrons in the core of a reactor.

Risks of thermal reactors include:
- release of radioactive material during operation
- release of radioactive material due to a fault
- the long half-life of radioactive waste, which means it must be kept safely away from the biosphere (living things) for 10 000 to 1 000 000 years.

To reduce the risks as much as possible, the reactors have thick concrete shielding to absorb radiation. There is a heat exchanger so that the coolant does not mix with the water in the power-generating plant. If radioactive dust or gas escaped, this would contaminate the environment and be dangerous to living things, so the reactor is enclosed and the level of radioactivity in the environment is monitored.

The design of the reactor is 'fail safe'. This means that, for example, the control rods can drop into the reactor to stop the reaction, and they do not rely on electric motors, in case the power switches off in an emergency.

Production of radioisotopes

In a fission reactor, radioisotopes can be made which do not exist naturally. These have a wide variety of uses.

Examples of uses of radioisotopes

Testing for cracks γ rays have the same properties as short-wavelength X-rays, so they can be used to photograph metals to reveal cracks. A γ source, such as cobalt-60, is compact and does not need an electrical power source like an X-ray tube.

Cancer treatment γ rays can penetrate deep into the body and kill living cells. So a highly concentrated beam from a cobalt-60 source can be used to kill cancer cells in a tumour. Treatment like this is called *radiotherapy*.

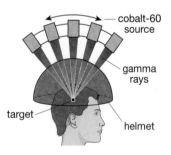

Tracers Radioisotopes can be detected in very small (and safe) quantities. This means that they can be used as tracers – their movements can be tracked. Examples include:
- tracking a plant's uptake of fertilizer from roots to leaves by adding a tracer to the soil water,
- detecting blocks in underground pipes by adding a tracer to the fluid in the pipe.

For imaging the thyroid gland, which takes up iodine, the radioisotope iodine-123 (γ-emitter with a half-life of 13 h) may be used. For treating thyroid cancer, iodine-131 (β- and γ-emitter with a half-life of 8 days) may be used.

To find a leak in a pipe, to check if packages are full, or to measure the thickness of materials, β-emitters are often the best choice. This is because α particles would always be absorbed before reaching the detector, and γ radiation would always pass straight through and be detected.

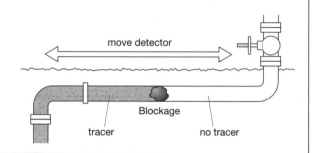

Carbon dating Living organisms are partly made from carbon which is recycled through their bodies and the atmosphere as they obtain food and respire. A tiny proportion is radioactive carbon-14 (half-life 5730 years). This is continually forming in the upper atmosphere as nitrogen-14 is bombarded by cosmic radiation. When an organism dies, no new carbon is taken in, so the proportion of carbon-14 is gradually reduced by radioactive decay. By measuring the activity, the age of the remains can be estimated to within 100 years. This method can be used to date organic materials such as wood and cloth. (See 'The decay law' on page 62 for a decay curve for carbon-14.)

Imaging the body (see Diagnosis methods in medicine on pages 72–74).

RTGs Radioisotope thermoelectric generators These are small power sources ideal for spacecraft that use energy from the radioactive decay of a radioactive isotope, for example plutonium-239.

Destructive uses of nuclear fission

The atomic bomb uses a chain reaction in a sample of radioactive material to release a large amount of energy. In addition to the effect of the heat and the blast radioactive isotopes are spread and contaminate a large area. Both uranium and plutonium bombs have been used (at Hiroshima and Nagasaki in Japan in 1945).

The plutonium produced in a nuclear reactor by nuclear fission can be used to produce a 'dirty bomb' in which fission does not occur but which spreads radioactive material over an area causing radioactive contamination.

Nuclear waste

There are three types of radioactive waste.

Low-level waste: for example, laboratory equipment and protective clothing. It can be sealed in drums and buried in trenches in clay rocks (so that if the drums corrode the radioactivity will not seep into the water supply).

Intermediate-level waste: for example, fuel cladding from nuclear reactors. The drums of waste can be put in concrete casks and then in reinforced concrete trenches. Clay above and below prevents contamination of water.

High-level waste: for example, spent fuel. This has to be stored in cooling water ponds until the level of activity drops enough for it to be sent for reprocessing. Liquid waste is vitrified (sealed in glass). The high-level waste is stored in steel canisters containing concrete. It can be kept in a safely guarded compound, or buried in deep underground sites.

The problem is whether the site will stay geologically stable for 1 000 000 years.

Do the benefits of thermal reactors outweigh the risks? To make this decision, the Government consults scientists for advice.

For some years no additional nuclear power stations were built because it was thought to be too risky. Now, to reduce our dependence on fossil fuel (because of global warming), there are plans to build more nuclear power stations. Newer designs are said to be safer, and we already buy electricity from France that is generated using nuclear reactors. Global warming could be more devastating than a nuclear reactor disaster, and the power cuts that would result from not having enough power-generating plants would be a risk to our health and well-being too. However, the problems of waste storage have not changed.

Irradiation and contamination

If the body was *irradiated* with α, β, and γ emissions:

- α particles would not penetrate the skin (they could cause skin cancer, but would not affect the internal organs)
- β radiation would penetrate the whole body and could cause damage to any organ
- γ radiation would mostly pass straight through the body without interacting with it.

If the body was *contaminated* with an α-, β-, or γ-emitting material (for example, by swallowing or breathing in dust or radon gas):

α particles are the most ionizing and would cause most damage (for example, lung cancer if breathed in).

Nuclear fusion

Reactors using *nuclear fusion* are many years away. Current research is based on the fusion of hydrogen-2 (called *deuterium*) and hydrogen-3 (called *tritium*):

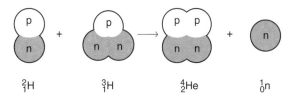

$$^2_1H \qquad ^3_1H \qquad ^4_2He \qquad ^1_0n$$

Although the energy release per fusion is less than 10% of that from a fission reaction, fusion is the better energy source if the processes are compared *per kg* of material.

Fusion is much more difficult to achieve than fission because the hydrogen nuclei repel each other.
- For the nuclei to have enough kinetic energy to overcome the electrostatic repulsion between them, and get close enough for the strong force to take effect (see page 57), the gas has to be heated to 10^8 K or more. At this temperature the atoms lose their electrons and the gas becomes a *plasma*.
- No ordinary container can hold such a hot material and keep it compressed. Scientists are experimenting with magnetic fields to trap the nuclei in a ring-shaped container (called a tokamak).

The advantages of a fusion reactor will be:
- Fuels will be readily obtainable. For example, deuterium can be extracted from sea-water.
- The main waste product, helium, is not radioactive.
- Fusion reactors have built-in safety. If the system fails, fusion stops.

The Sun gets its energy from the fusion of hydrogen, though using a different reaction from that on the left. Its huge size and gravity maintain the conditions needed.

Fusion in stars

Stars form in huge clouds of gas (mainly hydrogen) and dust called nebulae. Gravity pulls more and more of the gas and dust into a clump called a protostar. As the gas is attracted closer together the pressure and temperature increase. Eventually the pressure and temperature are great enough for fusion reactions to start.

Thermal activity stops gravitational collapse and the star has become a main sequence star. Our Sun is currently a main sequence star.

The Sun gets most of its energy from the proton-proton chain a multi stage process that converts hydrogen-1 into helium-4.

$$^1_1H + {}^1_1H \rightarrow {}^2_1H + {}^0_1e + \nu$$

$$^1_1H + {}^2_1H \rightarrow {}^3_2He + \gamma$$

$$^3_2He + {}^3_2He \rightarrow {}^4_2He + {}^1_1H + {}^1_1H$$

Once all the hydrogen is converted to helium fusion stops and the star cools, the inner core collapses and the temperature rises until the pressure and temperature are high enough for the helium to start to fuse. The outer regions expand, cool and turn redder. The star is now a red giant.

Energy released in nuclear reactions
Example

$$^3_2He + {}^3_2He \rightarrow {}^4_2He + {}^1_1H + {}^1_1H$$

On the left hand side, the energy is the rest mass of the two helium nuclei.

On the right hand side, the energy is the rest mass energy of the helium nuclei and the two protons (hydrogen nuclei).

The energy released is given by $\Delta E = \Delta m\, c^2$ where Δm is the difference in the mass.

nuclide	Rest mass / u
$^3_2$He	3.0160
$^4_2$He	4.0026
$^1_1$H	1.0078

$m = 2 \times 3.0160\ u - 4.0026\ u - 2 \times 1.0078\ u$

$\Delta m = (6.0321 - 6.0183)\ u = 0.0138\ u$

$\Delta E = 0.0138\ u \times 1.661 \times 10^{-27}\ kg\ u^{-1} \times (3.0 \times 10^8\ m\ s^{-1})^2$

$\Delta E = 2.063 \times 10^{-12}\ J = 2.1 \times 10^{-12}\ J\ (2sf)$

This can be expressed in MeV:

$\Delta E = 2.063 \times 10^{-12}\ J \div 1.6 \times 10^{-13}\ MeV\ J^{-1} = 13\ MeV$

Module 4: Medical imaging

5.4.1 X-rays

X-rays

X-rays are photons of electromagnetic radiation with wavelength ~ 10^{-10} m.

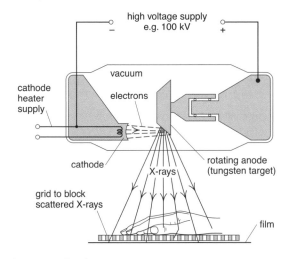

In the X-ray tube above:
- electrons are accelerated by the high p.d. across the tube gaining high energy E_k
- the electrons strike a metal target
- about 1% of the E_k is converted into X-ray photons; the rest is released as heat
- the anode is rotated rapidly to prevent 'hot spots'.

Energy and wavelength

- the X-ray beam contains a range of energies, and therefore, wavelengths (Energy = $hf = hc/\lambda$)
- the longer, least penetrating wavelengths are called soft, or low energy, X-rays
- the shorter, more penetrating ones are called hard, or high energy, X-rays
- low energy X-rays do not penetrate the skin, but are absorbed
- a filter, for example aluminium, can be used to absorb the low energy X-rays.

Tube current and voltage

- increasing the current increases the intensity of the beam
- increasing the voltage increases the intensity and reduces the peak wavelength, i.e. it produces more photons, which are more penetrating.

Intensity

> **Intensity** (in W m^{-2}) is power per unit cross sectional area

$$\text{intensity} = \frac{\text{power}}{\text{cross sectional area of surface}}$$

On the right, waves are radiating uniformly from a source of power output P. At a distance r from the source, the power is spread over an area $4\pi r^2$.

So intensity $I = \dfrac{P}{4\pi r^2}$

Note that $I \propto \dfrac{1}{r^2}$.

This is an example of an inverse square law.

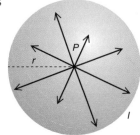

Interactions of X-rays with matter

There are three interactions that cause most of the attenuation of an X-ray beam in soft tissue.

When an X-ray photon strikes an electron in an atom, what happens will depend on the energy that is needed to remove the electron from the atom (the binding energy of the electron). The binding energy of the electron is much smaller for the outer orbital electrons than it is for those close to the

Energy of the X-ray photon	Interaction
Less than the binding energy of an orbital electron	Simple scattering – the photon is deflected without a change in energy.
Slightly greater than the binding energy of an orbital electron.	The photoelectric effect
Much greater than the binding energy of an orbital electron. Comparable to the rest mass of the electron (9.11×10^{31} kg equivalent to 0.51 MeV)	The Compton effect
More than twice that of the rest mass of the electron (i.e. greater than 1.02 MeV)	Pair production

Note: simple scattering is included for completeness – do not give this as an example of an interaction.

nucleus of the atom.

The photo-electric effect

All the energy of the photon is given to the electron which leaves the atom as a photoelectron with a large amount of kinetic energy. The photoelectron will collide with, and

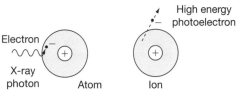

ionise, other atoms before all its kinetic energy is dissipated.

At AS level you learned about the photoelectric effect in metals: when photons strike a metal surface they may cause emission of free electrons as photoelectrons. The equation for the photoelectric effect is:

$hf = \Phi + \tfrac{1}{2}mv^2$

where f is the frequency of the photon, hf is its energy, Φ is the work function of the metal, which is the minimum energy needed to release a photoelectron, and $\tfrac{1}{2}mv^2$ is the maximum kinetic energy of the photoelectron emitted from the surface. (If the electron is not at the surface and requires more energy than Φ to release it, it will have less kinetic energy and so there will be a spread of kinetic energies).

Example of X-ray striking a metal surface:

If the X-ray wavelength is 1.0×10^{-10} m and the work function is 2.9 eV, using $f = c/\lambda$:

$6.63 \times 10^{-34} \text{Js} \times \dfrac{3.0 \times 10^8 \text{ m s}^{-1}}{1.0 \times 10^{-10} \text{ m}} \times \dfrac{1.0}{1.6 \times 10^{-19} \text{ J eV}^{-1}}$

$= 2.9 \text{ eV} + \tfrac{1}{2}mv^2$

$\tfrac{1}{2}mv^2 = 12000 \text{ eV} - 2.9 \text{ eV}$

So for X-ray photons the work function of the metal is negligible compared to the energy of the photon.

Interactions of X-rays with matter *continued*

For an atom in soft tissue:

$$hf = E_b + \tfrac{1}{2}mv^2$$

where E_b is the binding energy of the electron, and $\tfrac{1}{2}mv^2$ is the kinetic energy of the photoelectron.

The Compton effect

Only a fraction of the energy is given to the electron, which recoils and is known as the recoil electron. The rest of the energy is carried away by an X-ray photon at an angle θ to the original photon direction, so that there is conservation of momentum.

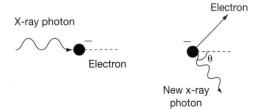

Pair production

The X-ray photon may interact with the nucleus of the atom. Its energy is absorbed and an electron and a positron are emitted in a process called pair production.

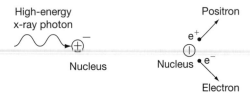

Note:

Using $\Delta E = \Delta m\, c^2$ gives

$$\Delta E = 9.11 \times 10^{-31}\ \text{kg} \times (3.0 \times 10^8\ \text{m s}^{-1})^2 = 8.2 \times 10^{-14}\ \text{J}$$

$$= \frac{8.2 \times 10^{-14}\ \text{J}}{1.6 \times 10^{-19}\ \text{C}} = 510000\ \text{eV} = 0.51\ \text{MeV}$$

The positron has the same magnitude mass as the electron so to create an electron-positron pair requires 2×0.51 MeV = 1.02 MeV energy.

This is more likely with gamma rays. Most X-ray machines used in medical diagnostics do not produce X-rays with 1 MeV energy.

Attenuation

As X-rays pass through a material, their energy is gradually absorbed and their intensity reduced. This is called **attenuation**. It is in addition to any reduction in intensity due to beam divergence.

When the X-ray beam is *collimated* (the rays are parallel), then the beam does not diverge. If I_0 is the intensity of the incident beam (do not confuse with I_0 as used in ultrasound), and I is the intensity at a distance x into the material, then, for a collimated beam,

$$I = I_0 e^{-\mu x} \tag{1}$$

μ is the **total linear attenuation coefficient**. Its value depends on the absorbing material and on the photon energy. X-rays are more attenuated by bone than by soft tissue. The difference is greatest for photons of around 30 keV energy.

The half-thickness $x_{\frac{1}{2}}$ of an absorber is the value of x for which $I/I_0 = \frac{1}{2}$. From equation (1): $x_{\frac{1}{2}} = \log_e 2/\mu$

X-ray photographs

X-rays affect photographic film, but cannot be focused, so the photos are 'shadow' pictures of the absorbing areas. For a sharp image, the X-rays need to be emitted as if from a point source. To reduce the risk of cell damage, exposure times are normally less than 0.2 s.

Image intensifiers

These use fluorescent screens which absorb the X-rays and re-emit visible light in the same pattern as the original X-rays.

Why they are used:
- most high energy X-ray photons pass straight through photographic film
- an image intensifier absorbs X-ray photons and emits many more visible light photons (do not confuse with the photoelectric effect which causes emission of electrons)
- lower exposure to X-rays is needed to produce an image, which is safer for the patient
- the visible light photons produce an image on photographic film, or for digital storage
- the pattern is the same as the original X-rays.

The main parts:
- X-rays enter through a window (often aluminium) that does not absorb or scatter X-rays
- a fluorescent screen, or phosphor, absorbs the X-rays and emits more light photons
- a photocathode emits electrons when light photons from the screen fall on it
- the electrons are in a vacuum and are accelerated towards another fluorescent screen or phosphor
- the electrons are absorbed and light photons from the screen travel towards the output window
- originally this image was stored on photographic film, but today it is usually stored as an electronic image.

Image contrast media

A contrast medium:
- has a high attenuation coefficient, or a high absorption coefficient
- is ingested or injected into the body
- improves the contrast between the tissues in the image
- the scan shows the outline, or shape, of the soft tissue.

Examples are:
- iodine-based which are injected
- barium sulphate, in a barium meal.

Patients drink the meal just before the X-ray is taken and this improves the contrast between the digestive system and the rest of the body.

CAT scanning

A simple X-ray is taken from only one direction. This produces a single 2D image.

In **CAT (Computerised Axial Tomography) scans**, also called **CT scans**:

- the body is scanned by an X-ray beam at lots of different angles (this is done by rotating a fan of narrow, monochromatic, X-ray beams around the body)
- the intensity reduction caused by each 'slice' is measured by a detector
- the data from the detector is processed by computer
- a 3D image of the body is constructed from all the slices
- the image is stored, and displayed on a screen.

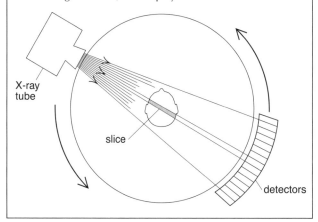

Advantages of CAT scans compared to X-ray images

- A simple X-ray produces a single 2D image – a CT scan produces a 3D image.
- If a simple X-ray is spread over a large area it will be blurred – a CT scan has better definition, and greater detail can be seen.
- CT scans have better contrast, and soft tissues can be seen.
- Sometimes it is not possible to position the simple X-ray source to get a clear line-of-sight of the area of interest, for example, a break in a bone may be obscured by other bones. CT scans clearly shows very small bones.
- In a CT scan the image can be rotated.

5.4.2 Diagnosis methods in medicine

Medical tracers
- Imaging techniques require good range and penetration, so γ-emitters are used.
- For imaging the body, the radioisotope technetium-99^m is useful. It is an γ-emitter with a half-life of 6 hours and will have effectively disappeared from the body in a couple of days.
- Another tracer used in medical diagnosis is the γ-emitting radioisotope iodine-123 (half-life 13 h). Small amounts can be carried in the bloodstream to various sites in the body.

Gamma camera
- γ photons from the tracer strike the sodium iodide disc, causing flashes of light.
- The light intensity is amplified by the photomultiplier tubes.
- Signals from these are processed electronically, and an image built up by computer.
- The collimator improves image quality by only letting through γ rays travelling at right angles to the disc.

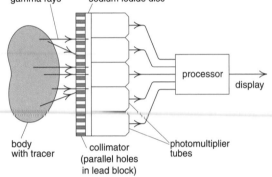

Checking blood flow in the lungs With technetium-99^m tracer present in the bloodstream, a gamma camera will reveal which parts of the lungs do not contain the tracer and, therefore, have blocked blood vessels.

Checking thyroid function For imaging the thyroid gland, which takes up iodine, the radioisotope iodine-123 (γ-emitter with a half-life of 13 h) may be used.

Positron emission tomography (PET)
- A radioisotope is used which emits positrons.
- It is injected or swallowed by the patient and accumulates in the part of the body being examined.
- When the radioisotope decays the positron is emitted but almost immediately collides with an electron, and both are annihilated with the release of two gamma photons of equal energy (0.51 MeV) moving in opposite directions (to conserve energy and momentum).
- The gamma photons are detected by the gamma camera. There will always be two and by comparing their time and position of arrival, the place where they originated can be determined.
- Another advantage is the all the other, single, gamma rays can be ignored, since the ones from the positron – electron interaction are always in pairs.

An example is the use of a compound similar to glucose which has a radioactive atom of fluorine-18 in it. This has a half-life of about 110 minutes and decays to oxygen-18 emitting a positron. This compound is taken up by the brain, which uses glucose, and it decays to form glucose. The scan shows areas which are very active, as they need more energy and so use more glucose. Cancerous tumours are also more active and need more energy than surrounding tissue so the same technique can be used to image tumours.

MRI (magnetic resonance imaging)
This produces 3D images of the body by scanning using magnetic fields.

Magnetic resonance in the body:
- About 60% of the human body is hydrogen atoms.
- The hydrogen nucleus - the proton - spins on its axis, so that it behaves as a small magnet.
- In a strong magnetic field a proton lines up parallel to the field and experiences a torque.
- The torque makes the spinning proton **precess** about the direction of the magnetic field (in a similar way to a spinning gyroscope precessing about the direction of the gravitational field).
- There are two states. Most protons line up in the direction of the field (the lower energy state) but some line up against the field (the higher energy state).

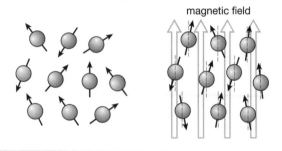

a) random protons

b) protons precessing around B field direction

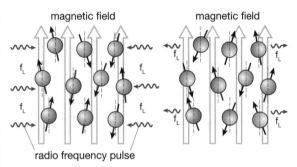

c) protons change energy state

d) relaxation

The Larmor frequency
The frequency of precession of the proton is called the **Larmor frequency** f_L, measured in Hz. It depends on the magnetic flux density B

$$f_L = 4.25 \times 10^7 B$$

- The larmor frequency is in the radio frequency (RF) range.
- When a pulse of electromagnetic radiation at this frequency is directed at protons some absorb the energy and move to the higher energy state.
- At other frequencies this does not happen – it is a resonance effect that occurs at frequency f_L.
- After the pulse the protons will relax back to the lower energy state, emitting RF waves that can be detected, amplified and recorded. (This is why a pulse, and not continuous, radio wave must be used.)
- These relaxation times are different for different tissue, for example, muscle, brain, and tumour tissue – this is the key to MRI.

MRI scanning

- The body is in a very strong, uniform, magnetic field.
- An additional, accurately calibrated, non-uniform magnetic field called a gradient field is used, so that every point in the body being scanned is in a different, known, magnetic field.
- This means that every point has a different, known, value of f_L, so that the part of the body (the position of the protons) the radio waves have come from is known from the frequency detected, f_L.
- The RF coils are used to vary the radio frequency f_l.
- The pulses of radio waves, which are emitted from different tissue with different relaxation times, are detected and sent to a computer.
- The computer processes the data to produce a map of the body showing different tissue types.
- False colour images are sometimes used, where a different brightness is assigned a different colour, to highlight different tissue.

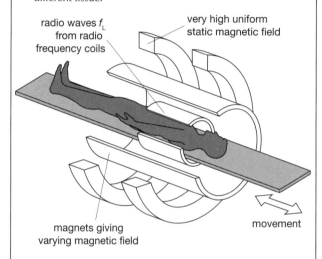

radio waves f_L from radio frequency coils

very high uniform static magnetic field

magnets giving varying magnetic field

movement

The MRI machine uses superconducting electromagnets, cooled by liquid helium, to produce a very strong magnetic field. Any ferromagnetic objects are strongly attracted to it. Staff and patients must be very careful not to take ferromagnetic metal objects into the room.

Uses of MRI scans

MRI scans can be used to:
- distinguish between cancerous tissue and normal tissue
- look for unusually shaped or sized organs
- look for damaged tissue, which often contains more water – the image is bright or dark depending on how much fat or water there is in the tissue (because of the different concentrations of hydrogen nuclei)
- look at arteries to see if they have become narrowed or are likely to rupture
- measure the signals in the brain and can see the changes due to neural activity. This can identify parts of the brain used for different tasks.

Advantages and disadvantages of MRI

Advantages:
- MRI scans do not use ionising radiation
- they give good contrast between different types of tissue
- they can be used for imaging the brain through the skull.

Disadvantages:
- MRI scanners are expensive
- they are slow
- no metal objects can be scanned (including metal pins and implants inside patients).

Non-invasive techniques in diagnosis

Non- invasive techniques are needed in diagnosis to be able to get information about the inside of the body without surgery.

The use of surgery would require a team of doctors and nurses, specialised equipment, operating equipment and wards for patients to recover.

If the initial exploratory surgery did not give the information expected or required, then further exploration might be needed, but there would be a limit on the amount of operations the patient could survive, and so a limit on the information that could be acquired.

Some surgery would be too damaging to attempt – for example in the brain, and some information would be inaccessible as it would require seeing under or inside other organs or bones.

After the surgery the patient would need some time to recover, requiring time off from school or work, whereas with non-invasive techniques they would recover in a few hours or less.

Non-invasive diagnosis can also be used with surgery, either before or during the operation, so that surgeons are prepared for what they find – for example by knowing the exact shape and position of a tumour.

The Doppler effect

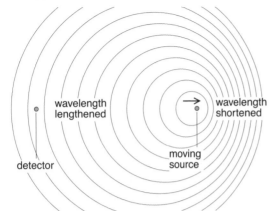

If a wave source is receding (moving away) from a detector, the waves reaching the detector are more spaced out, so their measured wavelength is increased and their frequency reduced. This is an example of the **Doppler effect**. It causes the change of pitch which you hear when an ambulance rushes past with its siren sounding.

$$\frac{\Delta f}{f} = -\frac{v}{c} \quad \text{and} \quad \frac{\Delta \lambda}{\lambda} = \frac{v}{c}$$

where v is the speed of the source and c is the wave speed, f and λ are the emitted frequency and wavelength. Δf and $\Delta \lambda$ are the observed changes – both defined as *increases*.

The same effect is observed if the source of the waves is stationary, and the observer is moving towards or away from the source. It is the relative motion between the source and the observer that causes the effect.

Doppler speed cameras

The speed of an object can be measured by sending a beam of waves towards it at detecting the change in frequency. The diagram shows how a laser speed camera send a beam towards a car, which reflects the beam at a higher frequency if the car is moving towards the camera. If the car is travelling faster there is a larger change in the frequency.

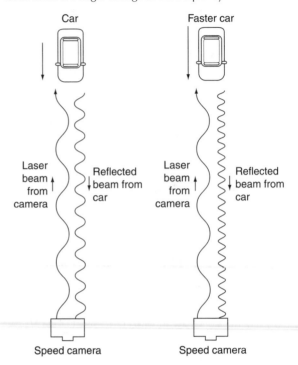

Note that just as when measuring the distance using a reflected beam there will be a factor of x2 in the frequency shift because the wave has been reflected. (The source is moving towards the observer while the beam is travelling towards it, and while it is returning.)

Answering questions

Re-read your answers and make sure that the examiner can understand what you mean.

When frequency increases, wavelength decreases, and vice versa. So writing: 'The Doppler effect is the change of wavelength and frequency when a source moves. If the source moves towards the observer it increases.' is not clear enough; the first sentence is OK, but the second could refer to frequency or wavelength so will gain no marks.

When you describe beams of travelling waves, some words can be used for distance or for time. Words such as 'longer' and 'shorter', for example, may be unclear.

'It will arrive at point B faster' is unlikely to gain an explanation mark, because 'It will arrive at point B faster **because it has taken a shorter route**' is quite different from 'It will arrive at point B faster **because it has a greater velocity**', and the examiner cannot tell whether the student understands the situation or not.

Blood flow

When a beam of ultrasound is sent into the body, any motion within the body causes a Doppler shift in the reflected ultrasound. This can be used, for example to check the heart beat or blood flow in an unborn baby. Any difference between the outgoing and returning frequencies is heard as a tone or displayed on a screen.

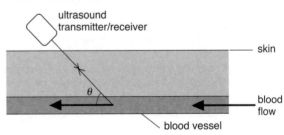

Although the ultrasound probe is to one side of the blood vessel, the component of the blood speed, $v \cos\theta$ can be found, and so v can be found.

5.4.3 Ultrasound

Ultrasound

Ultrasonic sound, or ultrasound, has frequencies above the upper limit of human hearing, i.e. above 20 Hz.

To take an ultrasound scan of part of the human body:
- a pulse of ultrasound is sent into the body
- the pulse is partially reflected at each boundary between different layers of tissue
- the time delay between the emitted and the reflected signal is used to determine the depth of the boundary
- the fraction of the signal reflected can be used to determine the type of tissue.

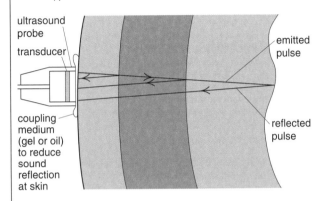

- The probe contains a transducer which sends and receives ultrasound pulses.
- The ultrasound is produced using the piezoelectric effect:
  - When a p.d. is applied across certain crystals they expand, or contract.
  - A high frequency alternating p.d. at the resonant frequency causes a slice of crystal to vibrate, emitting ultrasound.
  - The reflected ultrasound causes the crystal to vibrate and this produces a p.d. across the crystal – the detected signal pulse.

- The detected signal can be displayed on an oscilloscope or processed by a computer.
- Frequencies used are in the 2MHz -10MHz range.
- Increasing the frequency, so the wavelength is shorter, gives higher resolution but poorer penetration.

There are two types of scan:

A-scan:
- an A-scan is in one direction only
- it is used to measure a distance or a depth
- the reflected pulses are displayed as peaks on an oscilloscope
- positions along the time axis are a measure of distances into the body.

B-scan:
- a B-scan uses a number of sensors (or one sensor in different positions or angles)
- it is used to build up a 2D or 3D image
- the display is a 2D or 3D image on a computer screen.

Note: It may help to remember **A** scan – **A**mplitude changes, **B** scan – **B**rightness changes.

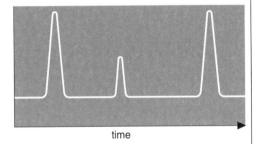

A-scan screen display

time

Destructive ultrasound Focused ultrasound, at an intensity above 10^7 W m^{-2}, can be used to break up kidney stones and gallstones, so surgery is not required.

Using ultrasound

Ultrasound is used to scan:
- the eye
- the unborn fetus
- other organs e.g. kidneys and gall bladder.

Advantages are:
- it is non-ionising radiation so does not damage DNA
- this means it is safer than X-ray imaging for patients and staff
- it is cheaper than MRI
- taking a scan is quicker than X-ray or MRI.

Disadvantages are:
- resolution is poorer than with X-ray imaging and MRI
- reflected strongly when passing from tissue to gas – so cannot be used to image the lungs
- specific impedance of bone is high, so it is not used for fractures. Images of the brain can be obtained , but MRI gives much better brain scans.

Measuring distance with ultrasound

Distance can be measured by sending out a pulse of waves and measuring the time it takes for the echo to return.

Other applications of this are ultrasonic range finders, ships' sonar, and radar.

If you know the speed of the waves, v, the distance, d, is

$$d = \frac{vt}{2}$$

It is important to remember to divide by 2, as the pulse has travelled there and back, twice the distance, in the time measured.

If d is known, this method can be used to work out the speed of the waves.

Ships' sonar uses the same technique to measure depth.

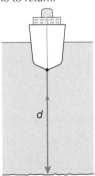

Acoustic impedance

A transmitting medium opposes the transmission of a sound pressure wave. The acoustic impedance is a measure of the resistance of a medium to the passage of sound waves.

The *specific impedance* of the medium $Z = \rho c$
where ρ is the density of the medium
and c is the speed of ultrasound in the medium.

Fraction of reflected intensity

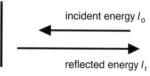

incident energy I_0

reflected energy I_r

When ultrasound reaches an interface between two different media the fraction of the incident energy that is reflected is called the fraction of reflected intensity and is given by

Fraction of reflected intensity $= \dfrac{I_r}{I_0} = \left(\dfrac{Z_2 - Z_1}{Z_2 + Z_1}\right)^2$

Specific impedances

Typical specific impedances are:

air	$430 \text{ kg m}^{-2} \text{ s}^{-1}$
fat	$1.38 \times 10^6 \text{ kg m}^{-2} \text{ s}^{-1}$
blood	$1.59 \times 10^6 \text{ kg m}^{-2} \text{ s}^{-1}$
muscle	$1.70 \times 10^6 \text{ kg m}^{-2} \text{ s}^{-1}$
bone	about $6 \times 10^6 \text{ kg m}^{-2} \text{ s}^{-1}$

Example

The ratio of reflected intensity to the initial intensity at the boundary between blood and muscle is:

$$\frac{(1.59 \times 10^6 - 1.70 \times 10^6)^2}{(1.59 \times 10^6 + 1.70 \times 10^6)^2} = 1.1 \times 10^{-3}$$

Acoustic impedance matching

When ultrasound crosses the air–skin boundary, almost all of it is reflected.

Acoustic impedance matching involves choosing a medium:
- with a similar, or identical, value of specific acoustic impedance, Z,
- so that little, or no, ultrasound is reflected at the boundary.

This is done by using a coupling gel which allows maximum transmission of ultrasound into the body.

Module 5: Modelling the Universe

5.5.1 The structure of the Universe

The contents of the Universe

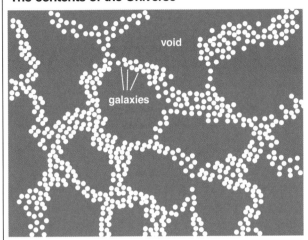

The Universe contains billions of galaxies. Their average separation is ~ 10^6 light-years. Together, they form a network of long, clumpy filaments with huge voids (spaces) in between. There are clusters of galaxies but, on a large scale, they are evenly distributed in all directions.

In addition to galaxies, the Universe contains:
• free hydrogen and helium
• neutrinos
• radiation

The current theory in cosmology is that 95% of the Universe is composed of dark matter and dark energy, however, these are not well understood at present.

The galaxies are clusters of stars, many of which have planetary systems like our solar system. The stars are at different stages of evolution ranging from the interstellar gas clouds which have yet to become stars to white dwarfs, neutron stars and black holes.

Solar System, stars, and galaxies

The Earth is one of seven or eight (depending on your opinions about Pluto) *planets* in orbit around the Sun. The Sun, planets, and other objects in orbit are together known as the *Solar System*.

Most of the planets move in near-circular orbits. Many have smaller planetary satellites (moons and artificial satellites) orbiting them. *Comets* are small, icy objects in highly elliptical orbits around the Sun. Planets, planetary satellites and comets are only visible because they reflect the Sun's light.

The Sun is one star in a huge star system called a *galaxy*. Our galaxy contains about 10^{11} stars, as well as interstellar matter (thinly-spread gas and dust between the stars). Our galaxy, called the *Milky Way*, is slowly rotating, with a period of more than 10^8 years. It is held together by gravitational attraction. It is just one of many billions of galaxies in the known *Universe*.

Normal galaxies emit mostly light. However, about 10% of galaxies have active centres which emit strongly in other parts of the electromagnetic spectrum as well.

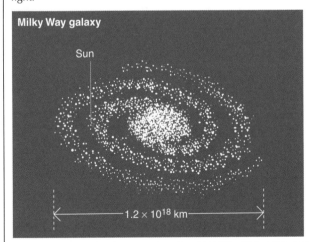

Milky Way galaxy

Sun

1.2×10^{18} km

Star formation

Stars form in huge clouds of gas (mainly hydrogen) and dust called *nebulae.* The Sun formed in a nebula about 5×10^9 years ago. The process took about 5×10^7 years:

Gravity pulled more and more nebular matter into a concentrated clump called a *protostar*. The loss of gravitational E_p caused a rise in core temperature which triggered the fusion of hydrogen and the release of energy. Thermal activity stopped further gravitational collapse. The Sun had become a main sequence star. (Its planets had formed in an orbiting disc of nebular matter.)

The future of the Sun

The Sun gets most of its energy from the **proton–proton chain**, a multi-stage fusion process which converts hydrogen-1 into helium-4.

The fusion process is:

$$^1_1H + {}^1_1H \rightarrow {}^2_1H + e+ + \nu + 0.42 \text{ MeV}$$

The positron immediately meets an electron and annihilates, releasing 2 gamma rays.

$$^2_1H + {}^1_1H \rightarrow {}^3_2He + \gamma + 5.49 \text{ MeV}$$

There are then several possible fusion reactions, but the most common one in the Sun is:

$$^3_2He + {}^3_2He \rightarrow {}^4_2He + 2{}^1_1H + 12.86 \text{ MeV}$$

Stars remain as main sequence stars until all the hydrogen in the core is fused to helium.

The Sun is about half way through its life on the main sequence (about 10^{10} years).

When all the hydrogen in the core has been converted to helium fusion will stop, and the core of the Sun will cool down. The core of the Sun will collapse again, under its own gravity, as it did during its formation.

This will cause heating of the outer part of the Sun until the hydrogen in the outer part is hot enough to fuse to helium. The outer part will expand and as it expands, the Sun will cool so that it becomes a **red giant**.

With the core temperature rising to over 10^8 K, energy is released by the fusion of helium into carbon.

After further changes, the outer layers expand and drift off into space. The core and inner layers become a ***white dwarf*** whose core is so dense that the normal atomic structure breaks down. The electrons form a ***degenerate electron gas*** whose pressure stops further collapse. Fusion ceases, and the white dwarf cools and fades for ever.

Note:
Stars less massive than about 0.5 times the Sun end up as white dwarfs, without going through the giant stages.

The future of massive stars

If a star has a mass greater than about 1.3 times the mass of the Sun, then it will be hotter and will fuse hydrogen to helium using the CNO cycle (a cycle in which carbon, nitrogen and oxygen take part) instead of the proton-proton cycle. Hotter, more massive stars consume hydrogen more quickly and are main sequence stars for a shorter time.

When all the hydrogen in the core has been converted to helium the core collapses, and if the star has a mass greater than 10 times the mass of the Sun it becomes a super **red giant**. The temperature and density of the core of a super red giant is enough to start the fusion of carbon, and when the carbon has fused the core collapses again and the increase in temperature starts the next fusion process. This cycle continues with heavier and heavier elements produced by fusion until the core of the star is iron. Iron is the most stable element so no further fusion reactions occur.

The core collapses again and this time there is a massive explosion called a **supernova**. These are very bright, and a lot of the material of the star is sent out with the shock wave that spreads out through space. During the explosion, neutrons combine with the nuclei of the heavier elements to form nuclei with masses larger than iron.

The small core remains. It is made mainly of neutrons and has a density similar to the nuclear matter in the nucleus (atoms are mainly empty space). If the mass is less than about twice that of the Sun it is a **neutron star**. If it has a higher mass then it becomes so dense that its gravitational field is too strong for light to escape and it is a **black hole**.

Distance units

In astrophysics, the following distance units are used.

Light-year (ly) This is the distance travelled (in a vacuum) by light in one year. 1 light-year = 9.47×10^{15} m.

Astronomical unit (AU) This is the mean radius of the Earth's orbit around the Sun. 1 AU = 1.50×10^{11} m.

Parsec (pc) This is the distance at which the mean radius of the Earth's orbit has an angular displacement of one arc second (1/3600 degree).

$$1pc = 3.26 \text{ ly} = 2.06 \times 10^5 \text{ AU} = 3.09 \times 10^{16} \text{ m}$$

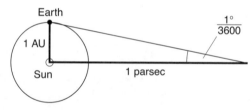

not to scale

diameter of Earth = 1.3×10^4 km
diameter of Sun = 1.4×10^6 km
radius of Earth's orbit = 1.5×10^8 km = 1 AU

diameter of Solar System = 50 AU
distance to nearest star = 2.7×10^5 AU = 4.2 ly = 1.3 pc
(*Proxima Centauri*)

diameter of galaxy (*Milky Way*) = 1.3×10^5 ly = 40 kpc
distance to neighbouring galaxy = 2.2×10^6 ly = 0.7 Mpc
(*Andromeda*)

Olbers' paradox In 1826, it was pointed out that, if the stars continued out to infinity, the night sky should be white, not dark – because light must be coming from every possible direction in the sky. This became known as Olbers' paradox.

Because the night sky is not white, Olbers' paradox can be resolved if the Universe is not infinite with an infinite number of stars.

Two reasons for the dark night sky have been suggested:
- In an expanding Universe, red-shifted wavelengths mean reduced photon energies, so the intensity of the light from distant stars is reduced.
- There is a limit to our observable Universe. If, say, the Universe is 15 billion years old, then we have yet to receive light from stars more than 15 billion light-years away. So everything beyond that distance looks dark.

Red shift

Information about a star's temperature composition and motion can be found by analysing its spectrum.

The frequency of the radiation from a star (or galaxy) is charged/shifted because of the star's motion. It is fast enough to cause a detectable Doppler shift in light waves. If a star is moving *away* from the Earth, its spectral lines are shifted towards the *red* end of the spectrum. If v is the relative velocity of recession, and v is small compared with the speed of light, c.

Example

The dark spectral lines from a galaxy are red shifted by 12%. What is its velocity?

$$\Delta\lambda = \frac{12 \times \lambda}{100} \text{ so using } \frac{\Delta\lambda}{\lambda} = \frac{v}{c}$$

$$v = \frac{(3 \times 10^8 \text{ m s}^{-1}) \times 12}{100} = 3.6 \times 10^7 \text{ m s}^{-1} \text{ away from Earth}$$
(because it is red shifted)

The Sun's rotation causes a broadening of its spectral lines, because light from the receding side is red-shifted while that from the approaching side is blue-shifted. This effect can be used to work out the speed of rotation.

The cosmological principle This says that, apart from small-scale irregularities, the Universe should appear the same from all points within it (i.e. the distribution of galaxies and their recession velocities should appear the same from all points).

Microwave background radiation

This is also called the cosmic microwave background radiation. It consists of microwaves that correspond to thermal radiation from an object with a temperature of about 3K. (For comparison, an object at 1000K glows red hot, and at 5000K it is white hot.) Thermal radiation has a range of wavelengths, and at 3K the peak intensity is a wavelength of about 1.9 mm which is in the microwave region. This radiation comes from all directions.

When the theory of the expansion of the Universe from a hot **big bang** was first suggested, scientists realised that the early Universe would have been white hot, filled with high energy gamma radiation which cooled as the Universe expanded. They predicted that today this radiation would be in the microwave region, corresponding to a very low temperature. This radiation would come from all directions.

In 1965 the radiation was discovered and is important evidence supporting the theory of the hot big bang model of the Universe.

Hubble's law

Edwin Hubble made many observations of the Doppler shifts of galaxies, and of their distance away from Earth. He found that all the distant galaxies had a red shift, which meant they were moving away from Earth, and the further away the galaxy the faster its speed away from Earth (speed of recession). This lead to:

> **Hubble's law** states that the speed of recession of a galaxy is proportional to its distance from the observer.

The fact that the speed is proportional to the distance means that all the distant galaxies are moving away from each other (not just from Earth) and the accepted explanation for this is that the Universe is expanding. This suggests that at some time in the past – at zero time – all matter and energy was together in a highly concentrated state. The age of the Universe is the time since the expansion began (known as the big bang, see 5.5.2 The evolution of the Universe, page 80)

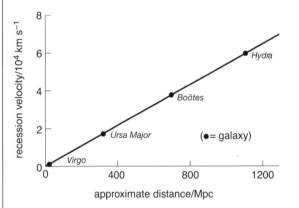

According to **Hubble's law**, the distance d of a galaxy and its recession velocity v are linked by this equation:

$$v = H_0 d \tag{1}$$

H_0 is called the **Hubble constant**.

It is usually quoted in units of 'kilometres per second per megaparsec' (km s^{-1} Mpc^{-1}) because the speed of the galaxy is in kilometres per second and the distance to the galaxy is in megaparsecs. If the units of speed are changed to m s^{-1} and distance to m, then the SI units of H_0 will be s^{-1}.

To change H_0 to SI units, if the estimated value of $H_0 = 65$ km s^{-1} Mpc^{-1} and 1pc $= 3.1 \times 10^{16}$ m:

$$H_0 = \frac{(65 \times 10^3 \text{ ms}^{-1})}{1 \times 10^6 \times (3.1 \times 10^{16} \text{ m})} = 2.10 \times 10^{-18} \text{ s}^{-1}$$

Large distances are difficult to estimate accurately, so the value of H_0 has a high uncertainty. However, it is thought to lie in the range 50–100 km s^{-1} Mpc^{-1} (1.6–3.2 $\times 10^{-18}$ s^{-1}).

Its value is important for several reasons:
- It enables the distances of the most remote galaxies to be estimated from their red shifts.
- The age of the Universe can be estimated from it (see page 80). (H_0 has dimensions of 1/time.)
- The fate of the Universe depends on it (see page 81).

5.5.2 The evolution of the Universe

The standard model of the Universe

This is the model which is currently accepted by most scientists. It is also known as the **hot big bang** theory.

According to this theory the Universe began with all matter and energy concentrated at a single point expanding in a burst of energy called the big bang. This was the moment at which time began – the hot big bang theory implies a finite age for the Universe. (Many previous models assume that the Universe has always existed.)

The ideas that time did not exist before the big bang (there is no 'before') and that space does not exist outside the Universe (it did not expand 'into' a space) most people agree are difficult to comprehend.

Current measurements indicate that the big bang was about 14 billion (13.7 x 109) years ago, or in other words, the age of the Universe is about 14 billion years old.

Estimating the age of the Universe

If a galaxy is d from our own, and has a steady recession velocity v, then separation of the galaxies must have occurred at a time d/v ago. This represents the approximate age of the Universe. From equation (1) $d/v = 1/H_0$, so

> age of the Universe $\approx 1/H_0$

This gives an age in the range 1–2×10^{10} years (10–20 billion years).

Note:
- The above calculation assumes constant v. However, recent observations suggest that the Universe's rate of expansion may actually be increasing, although the reason for this is not yet clear.

Time line after the big bang

For the first 10^{-43} s after the big bang little is understood. At the end of this time the temperature was about 10^{32}K and the four forces of gravitation, strong, weak and electromagnetic were united as just one force. After that the following sequence of events took place:

- At 10^{-43} s, gravity begins to behave as a separate interaction. The strong, weak and electromagnetic forces are still one force. This is sometimes called the Grand Unification Epoch. Quarks exist. (There is no distinction between quarks and leptons.) By the end of this time there may already have been different amounts of matter and antimatter in the Universe.

- 10^{-35} s after the big bang, when the temperature has fallen to 10^{27} K, the strong force becomes a separate interaction. The Universe expands very quickly until 10^{-32} s. There are now leptons and quarks.
- By about 10^{-12} s the weak force separates from the electromagnetic force.
- At about 10^{-6} s (1 microsecond after the big bang) quarks start to combine to form hadrons, so protons and neutrons are formed.
- Between 1s and 3 minutes neucleosynthesis occurred – the nuclei of light elements form (^{2}H, He, Li, Be). Matter is now 74% hydrogen 25% helium and 1% heavier nuclei.
- After about 1000 years the Universe is an expanding cooling gas of hydrogen and helium ions.

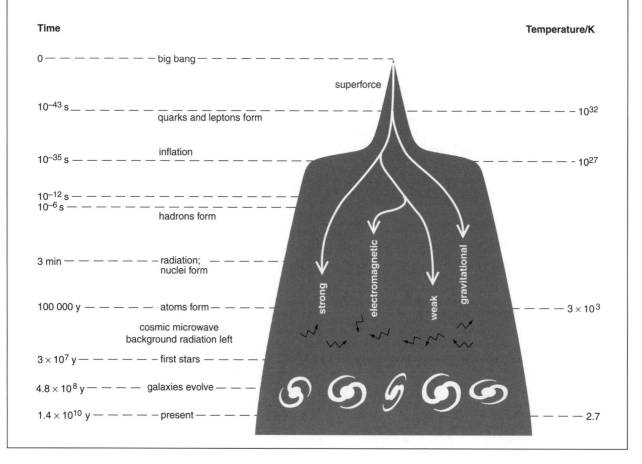

Time line after the big bang *continued*

- After about 100 000 years the temperature is 3000K electrons and protons combine into hydrogen atoms. Photons are no longer able to react freely with the atoms and the Universe becomes transparent. The photons have been travelling through the Universe ever since and they form the cosmic microwave background radiation which has been cooling ever since that time. The atoms are now neutral so electromagnetic repulsion is cancelled out and gravity begins to pull matter together. This is sometimes called the dark ages.
- According to more recent research, 30 million years after the big bang the first stars began to shine. (It used to be thought that this occurred 100 million years or even 200 million years after the big bang.)

The fate of the Universe

Gravity is affecting the expansion of the Universe. The fate of the Universe depends on how its average density ρ compares with a certain **critical density** ρ_0:

- If $\rho < \rho_0$ the expansion continues indefinitely.
- If $\rho = \rho_0$ the expansion continues, but the rate falls to zero after infinite time.
- If $\rho > \rho_0$ the expansion reaches a maximum, and is followed by contraction.

Linking ρ_0 and H_0 The critical density depends on the value of the Hubble constant. A higher H_0 means a higher recession velocity per unit separation. So a higher density is needed to stop the expansion. It can be shown that

$$\rho_0 = \frac{3H_0^2}{8\pi G}$$ where G is the gravitational constant

(see 4.2.2 Gravitational fields, page 26)

This gives ρ_0 in the range $5–20 \times 10^{-27}$ kg m^{-3}.

The average density of the Universe is currently believed to be close to the critical density. This would mean that the Universe is flat.

Models of the Universe

The Big Bang was not an explosion into existing space. Space itself started to expand. The galaxies are separating because the space between them is increasing.

Space has three dimensions of distance (represented by x, y, and z co-ordinates) and one of time. According to Einstein's theory of general relativity, gravity causes a curvature of space-time. If gravity is sufficiently strong, it may produce a 'closed' Universe, as shown below.

To visualize the expansion of the Universe, it is simpler to use models with only two of the distance dimensions. Imagine that the Universe is on an expanding, elastic surface. Three possible models are shown below. In each case, the galaxies move apart as the surface stretches. From any position on the surface, each galaxy recedes at a velocity that is proportional to its distance away.

Note:
- It is the value of the critical density, and therefore of the Hubble constant, which decides whether we live in an open, flat, or closed Universe.

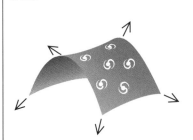

Open Universe ($\rho < \rho_0$) The surface is infinite and unbounded.

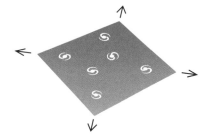

Flat Universe ($\rho = \rho_0$) The surface is infinite and unbounded.

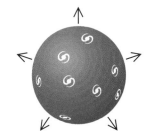

Closed Universe ($\rho > \rho_0$) The surface is finite and bounded.

Self-assessment questions

After revising a section you should try these questions.

Questions are only given for those sections which relate to compulsory material in the specifications.

Answers, including references to sections where you can find more detail, begin on page 84.

You should review the work relating to any questions that you were unable to do or when you obtain incorrect numerical answers.

Where necessary assume that the acceleration of free fall $g = 9.81$ m s^{-2}.

Unit 4 module 1

1. State the similarities and differences between an *elastic* and an *inelastic* collision.
2. A truck of mass 5000 kg travelling at 30 m s^{-1} collides with the rear of a car of mass 800 kg travelling in the same direction at 20 m s^{-1}. The two vehicles move together after the collision. Calculate the final velocity.
3. A nucleus of mass 4.00×10^{-25} kg splits up into two parts. One of mass 0.068×10^{-25} kg moves off at a speed of 1.0×10^7 m s^{-1}. Calculate the speed of the other particle.

Unit 4 module 2

1. (a) State what is meant by *angular frequency*.
   (b) State the unit in which it is measured.
2. A disc of radius 0.080 m rotates at a rate of 1500 revolutions per minute. Calculate:
   (a) the period of rotation of the disc
   (b) the speed of the rim of the disc.
3. A proton of mass 1.7×10^{-27} kg and speed 2.0×10^6 m s^{-1} moves in a circular path of radius 0.40 m. Determine the magnitude of the force acting on the proton. State the direction of this force at any instant.
4. Explain why a body moving at constant speed in a circular path is accelerating.
5. How is the centripetal force provided in the following situations?
   (a) A cyclist moving round a bend.
   (b) A car going round a bend on a very slippery banked track.
   (c) An orbiting satellite.
6. Draw two cycles of a displacement–time graph for SHM. Show on your diagram the amplitude and period of the oscillation.
7. A particle is released and moves with SHM of frequency 0.5 Hz and amplitude 0.040 m.
   Calculate:
   (a) the period of the motion
   (b) the maximum acceleration of the particle
   (c) the maximum velocity of the particle.
8. The period of a simple pendulum is 1.2 s. It swings with a maximum displacement of 0.018 m. The mass of the bob is 0.050 kg. Calculate:
   (a) the total energy of the pendulum
   (b) the kinetic energy when the bob is at half its maximum displacement
   (c) the potential energy in this position.
9. (a) Explain what is meant by a forced oscillation.
   (b) Under what conditions does resonance occur?
10. Mars has a mass of 6.4×10^{23} kg and a radius of 3.4×10^6 m. Determine the acceleration of free fall at its surface.
11. The acceleration of free fall at the surface of the Moon is 1/6 that at the surface of the Earth. Calculate the speed and period of a satellite in orbit close to the surface of the Moon. Radius of the Moon is 1.7×10^6 m.

Unit 4 module 3

1. A cook forgets about a saucepan on a cooker. The pan initially contains 0.250 kg of boiling water. The power supplied to the saucepan from the cooker is 0.600 kW. How long will it be before the saucepan boils dry? (Specific latent heat of water = 2.3×10^6 J kg^{-1}.)
2. State the difference between evaporation and boiling.
3. (a) State Boyle's law.
   (b) Explain what is meant by an *ideal gas*.
4. The temperature of a fixed mass of an ideal gas changes from 27 °C to 87 °C at constant volume. The original pressure was 0.9×10^5 Pa. Calculate the final pressure.
5. Calculate the final pressure for the same temperature change and initial pressure as in question **4** but allowing the volume of the gas to increase by 50%.
6. Calculate the number of moles of gas in a container of volume 2.5×10^{-3} m^3 containing gas at a pressure of 5×10^4 Pa at a temperature of 300 K.
7. Calculate the kinetic energy of a molecule at a temperature of 300 K. The Boltzmann constant is 1.38×10^{-23} J K^{-1}.
8. Determine the root mean square speed of oxygen molecules at room temperature, if the root mean square speed of hydrogen molecules is 1800 ms^{-1}.

   molar mass hydrogen = 0.002 kg mol^{-1}
   molar mass oxygen = 0.032 kg mol^{-1}

Unit 5 module 1

1. State two similarities and one difference between electric and gravitational fields.
2. Calculate the force between two point charges of magnitude 3.2×10^{-19} C and 1.6×10^{-19} C separated by a distance of 2.5×10^{-10} m.
3. An electron of charge -1.6×10^{-19} C is placed in a uniform field of strength 20000 V m^{-1}. Calculate the force on the electron.
4. (a) The unit of magnetic flux density is the Tesla. Define the Tesla.
   (b) Calculate the force on a straight wire of length 0.20 m when it carries a current of 2.5 A in a magnetic field of flux density 50 mT.
5. Explain the operation of a brake that depends on electromagnetic induction.
6. The flux density through a 200 turn coil of area 8.5×10^{-4} m^2 changes from 0.03 T to 0.12 T in 15 ms. Calculate the induced e.m.f. in the coil.
7. Calculate the magnitude of the force on an electron moving at a speed of 1.5×10^6 m s^{-1} in a magnetic field of flux density 0.20 T.
8. An electron starts travelling perpendicular to a uniform field. Explain why its path is circular if the field is magnetic, and parabolic if the field is electric.
9. Calculate the speed of an electron when it is accelerated through a potential difference of 2.5 kV.
10. In a mass spectrometer a designer wants to select ions with a velocity of 1.2×10^7 m s^{-1}. The magnetic field available has a flux density of 0.8 mT. Determine the strength of the electric field required to select this velocity.

Unit 5 module 2

1. State what is meant by a capacitance of 220 μF.
2. A 100 μF capacitor is charged to a p.d. of 6.0 V. It then discharges through a 2.2 kΩ resistor.
   How long will it take for the voltage to (a) halve (b) fall to 2.0 V?
3. A 100 μF capacitor is in series with a 200 μF capacitor. The p.d. across the combination is 12 V. Calculate
   (a) the total capacitance of the combination
   (b) the energy stored.

Unit 5 module 3

1. Write down the symbols for the following particles including their mass and charge numbers.
   Proton, electron, alpha particle, neutron, gamma ray photon
2. Describe the atomic structure of $^{206}_{82}Pb$.
3. State what is meant when we say that two nuclides are isotopes.
4. $^{206}_{82}Pb$ is formed when a radioactive atom decays by alpha emission. Determine the proton number and mass number of the atom that has decayed.
5. Complete the nuclear equation for the fusion reaction below and identify the particle X.

   $$4^1_1p \rightarrow {}^4_2He + 2X + 2v_e + \gamma$$

6. A small source emits gamma rays. The count rate is 280 s^{-1} when the GM tube is 10 cm from the source. What would you expect it to be when the tube is
   (i) 20 cm from the source
   (ii) 15 cm from the source.
7. Thorium-234 has a half-life of 24 days. A sample initially contains 6.0×10^{12} atoms. Calculate
   (a) the decay constant of thorium in s^{-1}
   (b) the initial activity of the thorium-234
   (c) the activity after 12 days.
8. Describe briefly one industrial and one medical use of radioactive materials. State the emission and the half-life of the source that is most appropriate for the application you choose.
9. In the fusion reaction in question 5 the proton has a mass 1.007 276 u, the alpha particle has a mass 4.001 506 u, and X has a mass 0.000 548 580 u.
   (i) Determine the mass defect in u.
   (ii) Calculate the energy in J liberated by the reaction.
   (1 u = 931 MeV; 1 MeV = 1.6×10^{-13} J.)
10. State the purpose in a nuclear reactor of (a) the moderator and (b) the coolant.
11. The rest energy of an electron is 0.511 MeV. Calculate its rest mass. (1 eV = 1.6×10^{-19} J.)
12. How much energy is released when an electron and a positron (each of mass 9.1×10^{-31} kg) annihilate each other? What form will the energy take?

13. An electron collides inelastically with an atom. Explain what is meant by an inelastic collision and why a collision with an atom may be inelastic.
14. Calculate the radius of a uranium-238 nucleus given that $R_0 = 1.2 \times 10^{-15}$ m.
15. To what class of particles do the following particles belong?
   electron-neutrino; neutron; electron
16. Write down the quark structure of an antiproton.
17. Use conservation laws to determine which of the following are possible reactions:
   (a) $\pi^+ \rightarrow \mu^+ + v_\mu$
   (b) $p^+ + \pi^- \rightarrow \pi^+ + K^-$

Unit 5 module 4

1. Ultrasound is used to examine an eyeball. The speed of ultrasound in the eye tissue is 1.5×10^3 ms^{-1}.
   (a) Explain why a gel is used between the probe and the eyeball.
   (b) The piezoelectric crystal transmits and receives the ultrasound. It emits short pulses of waves. Explain why there is a time interval between the pulses.
   (c) A pulse reflected from the back of the eyeball is detected 3.2×10^{-5} s after emission. Calculate the depth of the eyeball.
2. An X-ray source has an attenuation coefficient of 50 m^{-1} for bone and 6.7 m^{-1} for soft tissue. The initial intensity of the X-rays is 5×10^3 Wm^{-2}. Calculate the intensity after passing through
   (a) 2.5 cm of bone
   (b) 9.0 cm of soft tissue.
3. The magnetic field in an MRI scanner is 1.20 T.
   (a) Calculate the Larmor frequency.
   (b) The magnetic field is varied linearly along the length of the patient. What effect does this have on the Larmor frequency? Explain the advantage of this variation in field.

Unit 5 module 5

1. A galaxy is travelling at 6×10^6 ms^{-1} and is 4×10^{24} m from Earth.
   (a) Calculate a value for the Hubble constant.
   (b) What value does this suggest for the age of the Universe?
2. The wavelength of the Sodium D line (λ = 589 nm) from a distant galaxy is shifted to 591 nm.
   (a) Is the galaxy moving towards us or away from us?
   (b) How fast is it moving relative to the Earth?
3. Observation of the spectrum from a distant star shows that the light on one side is shifted to longer wavelengths and on the other side to higher wavelengths. What does this tell you about the star?

Self-assessment answers

Unit 4 module 1

1. Total energy and momentum are conserved in both types. In an elastic collision kinetic energy is conserved.
2. $28.6\,\mathrm{m\,s^{-1}}$ in original direction of motion.
3. $1.7 \times 10^5\,\mathrm{m\,s^{-1}}$

Unit 4 module 2

1. (a) The angle swept out by a radial line per second.
   (b) $\mathrm{rad\,s^{-1}}$ (radian per second)
2. (a) $0.040\,\mathrm{s}$
   (b) $13\,\mathrm{m\,s^{-1}}$
3. $1.7 \times 10^{-14}\,\mathrm{N}$ toward the centre of the circular orbit.
4. Velocity is a vector, so when the direction changes, as it does in circular motion, the velocity changes. There is acceleration because this is rate of change of velocity.
5. (a) The friction between the tyre and the road.
   (b) The horizontal component of the normal reaction to the track.
   (c) For an Earth satellite, the gravitational force between the Earth and the satellite.
6.

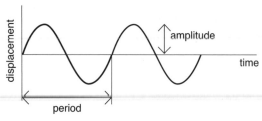

7. (a) $0.20\,\mathrm{s}$ (b) $39\,\mathrm{m\,s^{-2}}$ (c) $1.3\,\mathrm{m\,s^{-1}}$
8. (a) $2.2 \times 10^{-4}\,\mathrm{J}$ $(A\omega = \text{maximum } v; \omega = 2\pi/T; E_K = \frac{1}{2}mv^2)$
   (b) $1.7 \times 10^{-4}\,\mathrm{J}$ (c) $0.55 \times 10^{-4}\,\mathrm{J}$
   (At half maximum displacement E_p is $\frac{1}{4}$ maximum and E_K is $\frac{3}{4}$ maximum.)
9. See page 34.
10. $3.7\,\mathrm{m\,s^{-2}}$.
11. Speed = $1700\,\mathrm{m\,s^{-1}}$.
    period = $6400\,\mathrm{s}$

Unit 4 module 3

1. $960\,\mathrm{s}$
2. See page 36
3. See page 39
4. $1.08 \times 10^5\,\mathrm{Pa}$
5. $0.72 \times 10^5\,\mathrm{Pa}$
6. $0.050\,\mathrm{mol}$
7. 6.2×10^{-21}
8. $450\,\mathrm{m\,s^{-1}}$

Unit 5 module 1

1. Similarities: for point charges or masses
   Both obey inverse square law for variation of force with distance: both obey inverse r law for variation of potential with distance; both act at a distance with no requirement for a material medium.
   Difference: electric fields may produce attraction or repulsion; gravitational fields only produce attraction.
2. $7.4 \times 10^{-9}\,\mathrm{N}$
3. $3.2 \times 10^{-15}\,\mathrm{N}$ in opposite direction to the direction of the field.
4. (a) The strength of the magnetic field when force of 1 N is exerted on a wire of length 1 m when it caries a current of 1 A.
   (b) $25\,\mathrm{mN}$

5. See page 49
6. $1.0\,\mathrm{V}$
7. $4.8 \times 10^{-14}\,\mathrm{N}$
8. When an electron moves at right angles to a uniform magnetic field the magnetic force is always perpendicular to the direction of motion of the electron and has constant magnitude. This is the condition for circular motion. An electron that starts at right angles to a uniform electric field is accelerated in the direction of the field and has constant velocity perpendicular to the field. This leads to a parabolic path.
9. $3.0 \times 10^7\,\mathrm{m\,s^{-1}}$
10. $9600\,\mathrm{N\,C^{-1}}$

Unit 5 module 2

1. $220\,\mathrm{\mu C}$ of charge is 'stored' for each volt of potential difference between the terminals of the capacitor.
2. (a) $0.15\,\mathrm{s}$ (b) $0.24\,\mathrm{s}$
3. (a) $67\,\mathrm{\mu F}$ (b) $4.8 \times 10^{-3}\,\mathrm{J}$

Unit 5 module 3

1. (a) $^1_1\mathrm{p}$ $^0_{-1}\mathrm{e}$ $^4_2\mathrm{He}$ $^1_0\mathrm{n}$ $^0_0\gamma$
2. 82 protons and 124 neutrons in the nucleus with 82 electrons 'orbiting' the nucleus
3. The nuclides have the same number of protons in the nucleus but a different number of neutrons.
4. Proton number 84; mass (nucleon) number 210
5. $4^1_1\mathrm{p} \rightarrow {}^4_2\mathrm{He} + 2^0_1\mathrm{X} + 2^0_0\nu_e + {}^0_0\gamma$; X is a positron.
6. (i) 70 (ii) 120
7. (a) $3.3 \times 10^{-7}\,\mathrm{s}$ (b) $2.0 \times 10^6\,\mathrm{Bq}$
   (c) $1.4 \times 10^6\,\mathrm{Bq}$
8. See page 66
9. (i) $0.026\,501\mathrm{u}$ (ii) $4.0 \times 10^{-12}\,\mathrm{J}$
10. See page 65
11. $9.1 \times 10^{-31}\,\mathrm{kg}$
12. $1.6 \times 10^{-13}\,\mathrm{J}$; 2 gamma rays
13. In an inelastic collision E_k is not conserved. The energy may result in ioinisation or excitation of the atom.
14. $7.4 \times 10^{-15}\,\mathrm{m}$
15. lepton; baryon; lepton
    (baryons are hadrons)
16. $\mathrm{u\bar{u}\bar{d}}$
17. (a) This is possible. Lepton and charge numbers are conserved.
    (b) This is not possible. Charge is conserved but baryon number is not. p is a baryon. All the others are mesons.

Unit 5 module 4

1. (a) Reduces the amount reflected and maximizes transmitted energy.
   (b) To allow reflected waves to return and be detected without interference with transmitted waves.
   (c) $2.4\,\mathrm{cm}$
2. (a) $1.4 \times 10^3\,\mathrm{W\,m^{-2}}$ (b) $2.7 \times 10^3\,\mathrm{W\,m^{-2}}$
3. (a) $51\,\mathrm{MHz}$ or $5.1 \times 10^7\,\mathrm{Hz}$
   (b) It will change the Larmor frequency. The location of the pulse along the body is known from the value of the Larmor frequency.

Unit 5 module 5

1. (a) $1.5 \times 10^{-18}\,\mathrm{s^{-1}}$
   (b) $6.67 \times 10^{17}\,\mathrm{s}$ (2.1×10^{10} years)
2. (a) away (b) $1.0 \times 10^6\,\mathrm{m\,s^{-1}}$
3. It is rotating (one side towards the Earth, the other side away from it).

Index

radioactive decay 59, 61, 62
radioactive contamination 13, 65
radioactive particles 59, 62
radioactive sources (activity) 59, 60, 64, 65
radioisotopes 64, 70
radiotherapy 64
rate of rotation 21
recoiling particles 19
red shift 76, 77
relativity 62, 79
resistance 16, 48, 50, 61
    in sound waves 28, 30, 31, 74
resonance 32, 70
right-hand grip rule 43
RMS current see root mean square current
rockets 17
root mean square current 38
rotation 21, 27, 47, 77
Rutherford, Ernest 11, 12, 53

S
safety features, vehicles 13, 19
satellite 26, 27, 75
satellite navigation systems 13
scalars 18
scientific notation 9
scientific theories 11, 12, 14, 26, 33, 37, 78, 79
seat belts 19
sensors 28, 73
SHM see simple harmonic motion
'shock absorbers' (dampers) 31

simple harmonic motion 28–30
smoke detectors 59
Solar System 75
solids 33, 34, 36
sound 72
    waves 74
space travel 14
specific heat capacity 36
specific latent heat of fusion 36
spectral lines 77
spectrum 75, 77
speed 16–18, 21, 22, 26, 33, 34, 38, 42–45, 62, 63, 69, 72, 74, 75
    cameras 72
    of light 62
springs 18, 28, 30, 31
stability 58, 62
standard form of numbers 9
stars 55, 58, 66, 75, 76
    formation 75
    neutron 55
strain energy 31
strong nuclear force 53, 55, 62
structure of the atom 53–6
Sun 24, 26, 65, 66, 75, 76

T
temperature 33–38, 66, 75–78
    scales 35
    thermodynamic 35
theories, scientific 11, 12, 14, 26, 33, 37, 78, 79
thermal energy 35, 36
thermal equilibrium 35

thermal radiation 77
thermal reactors 63, 64
thermodynamic temperatures 35
torque 70
tracers 64, 70
transformers 52
transmutation 56

U
ultrasound 68, 72–4
ultraviolet radiation 13
uncertainties 9, 10
Universe 58, 75, 76–9
    models of 78, 79
uranium 54, 63

V
vectors 18, 19, 23
vehicle safety features 19
velocity 19, 21, 22
    in a circle 21
vernier scales, reading 10
voltage 48, 52, 61, 67

W
wavelengths 60, 64, 67–69, 73
weight 16, 17, 23, 25, 30
'weightlessness' 26
work 19, 31, 41, 48, 49, 62, 77
work function 67

X
X-rays 56, 64, 67–9, 73